T0360655

Nuclear Physics and Gamma-Ray Sources for Nuclear Security and Nonproliferation

Proceedings of the International Symposium

Nuclear Physics and Gamma-Ray Sources for Nuclear Security and Nonproliferation

Ricotti, Tokai-mura, Japan   28–30 January 2014

Editors

Takehito Hayakawa
Masao Senzaki
Paul Bolton
Ryoichi Hajima
Michio Seya

Japan Atomic Energy Agency

Mamoru Fujiwara

Osaka University

 **World Scientific**

NEW JERSEY · LONDON · SINGAPORE · BEIJING · SHANGHAI · HONG KONG · TAIPEI · CHENNAI

Published by

World Scientific Publishing Co. Pte. Ltd.

5 Toh Tuck Link, Singapore 596224

USA office: 27 Warren Street, Suite 401-402, Hackensack, NJ 07601

UK office: 57 Shelton Street, Covent Garden, London WC2H 9HE

British Library Cataloguing-in-Publication Data
A catalogue record for this book is available from the British Library.

NUCLEAR PHYSICS AND GAMMA-RAY SOURCES FOR NUCLEAR SECURITY
AND NONPROLIFERATION
Proceedings of the International Symposium Nuclear Physics and Gamma-Ray Sources for Nuclear Security and Nonproliferation

ISBN 978-981-4635-44-8

In-house Editor: Rhaimie Wahap

Typeset by Stallion Press
Email: enquiries@stallionpress.com

Printed in Singapore

Proceedings of the International Symposium

Nuclear Physics and Gamma-ray Sources for Nuclear Security and Nonproliferation (NPNSNP)

January 28-30, 2014, Ricotti, Tokai-mura, Japan

Editors
Takehito Hayakawa
Masao Senzaki
Paul Bolton
Ryoichi Hajima
Michio Seya
Japan Atomic Energy Agency

Mamoru Fujiwara
Osaka University

Hosted by
Integrated Support Center for Nuclear Nonproliferation and Nuclear Security
(ISCN) & Quantum Beam Science Directorate (QuBS), Japan Atomic Energy
Agency (JAEA)

vi

Organization of NPNSNP

[Invited Speakers]

Bernhard Ludewigt (Lawrence Berkeley National Laboratory)
Calvin R. Howell (Duke University)
CharyRangacharyulu (University of Saskatchewan)
Christopher P.J. Barty (Lawrence Livermore National Laboratory)
DenizSavran (Extreme Matter Institute (EMMI) / GSI)
Dino Jaroszynski (University of Strathclyde)
HaruoMiyadera (Toshiba Corporation)
Hideaki Ohgaki (Kyoto University)
Hiroaki Utsunomiya (Konan University)
Junji Urakawa (KEK)
Laurent Audouin (Centre National de la Recherche Scientifique)
Luca Serafini (INFN-Milan)
Masayuki Igashira (Tokyo Institute of Technology)
Massimo Ferrario (INFN-Milan)
Mitsuru Uesaka (University of Tokyo)
Robert Ledoux (Passport System Inc.)
Ronald Schwengner (Helmholtz-Zentrum Dresden-Rossendorf)
Shuji Miyamoto (University of Hyogo)
Sydney Gales (ELI-NP)
Vladimir Nedorezov (Institute for Nuclear Research RAS)

[International Advisory Committee]

Christopher P.J. Barty (Lawrence Livermore National Laboratory)
Luca Serafini (Istituto Nazionale di Fisica Nucleare, Sezione di Milano)
Mitsuru Uesaka (University of Tokyo)
Muhsin N. Harakeh (KernfysischVersneller Instituut, University of Groningen)
Shigeru Kubono (RIKEN)
Sydney Gales (ELI-NP)
William Bertozzi (Massachusetts Institute of Technology)
Yoshiaki Kiyanagi (Hokkaido University)
Yoshihiro Nakagome (Japan Nuclear Energy Safety Organization)

[Local Organizing Committee]

Masao Senzaki (Chair), Paul Bolton (Co-chair), Mamoru Fujiwara, Takehito Hayakawa, Ryoichi Hajima, Toshiyuki Shizuma, Michio Seya, Yosuke Naoi, Naoki Kobayashi

Preface

The international symposium "Nuclear Physics and Gamma-ray Sources for Nuclear Security and Nonproliferation" (NPNSNP) was held in Tokai-mura in Japanduring January 28th–30th, 2014. The event was hosted by the Integrated Support Center for Nuclear Nonproliferation and Nuclear Security (ISCN) and the Quantum Beam Science Directorate (QuBS), both relevant parts of the Japan Atomic Energy Agency (JAEA). ISCN was established as a result of the first nuclear security summit in 2010 which was held in Washington, D.C. QuBS researchers conduct research and development of new quantum beam sources with an assortment of applications in mind.

Nuclear nonproliferation is a critical global issue. A key technological challenge to ensuring nuclear nonproliferation and security is the detection of long-lived radioisotopes and fissionable nuclides in a non-destructive manner. This challenge requires new methods for detecting relevant nuclides and associated development of new quantum-beam sources. For example, one new method that is being examined is nuclear resonance fluorescence with energy-tunable, monochromatic gamma-rays generated by Compton scattering of laser photons with energetic electrons. The development of new methods requires expertise from a wide range of fields, such as nuclear physics, accelerator physics, laser physics, etc. Furthermore, new methods must be compatible with the requirements of administrators and nuclear-material inspectors. The purpose of this symposium was to provide a valuable forum for information exchange between researchers from these different fields.

This symposium attracted about 70 participants, of whom about 20 were from outside of Japan. It fostered many discussions between researchers representing a wide range of fields. Researchers associated with fundamental science realize that their work can contribute to nuclear security and correlated topics, whereas those involved in nuclear security issues are aware of much help made available by many scientists. Clear interest was expressed in establishing an ongoingsymposium series. Consequently, in 2016 the next symposium will be held at the Lawrence Livermore National Laboratory in California. We sincerely thank all participants and supporting staff members for making this unique gatheringa great success.

T. Hayakawa, M. Senzaki, P. R. Bolton,
R. Hajima, M. Seya, M. Fujiwara

CONTENTS

Preface.. ix

Oral Presentations 1

Nuclear Science and applications with the Next Generation of High-Power
Lasers and Brilliant Low-Energy Gamma Beams at ELI-NP
S. Gales 3

New Non-Intrusive Inspection Technologies for Nuclear Security and
Nonproliferation
R. J. Ledoux 15

Development of Gamma-Ray Nondestructive Detection and Assay Systems
for Nuclear Safeguards and Security at JAEA
R. Hajima.. 25

Photonuclear Reaction Studies at HIγS: Developing the Science of Remote
Detection of Nuclear Materials
C. R. Howell.. 33

Dipole Strength Around the Particle Threshold
D. Savran.. 43

Resonant Photonuclear Reactions for Neutrino Nuclear responses and
Nuclear Isotope Detections
H. Ejiri.. 49

Non-Destructive Testing for Control of Radioactive Waste Package
S. Plumeri and F. Carrel.. 55

Development of First Responders Equipment at RN Incident Sites
K. Tsuchiya *et al.* 59

Compact Short-Pulsed Electron Linac Based Neutron Sources
for Precise Nuclear Material Analysis
M. Uesaka *et al.* 67

Laser-Driven Plasma Deceleration of Electron Beams for Compact
Photon Sources
J.-L. Vay et al... 77

Advanced Laser-Compton Gamma-Ray Sources for Nuclear Materials
Detection, Assay and Imaging
C. P. J. Barty 85

Compact Gamma-Beam Source for Nuclear Security Technologies
P. Gladkikh and J. Urakawa... 97

Developments of Optical Resonators and Optical Recirculators for
Compton X/γ Ray Machines
A. Martens.. 107

Low-lying "Pygmy" Dipole Resonances and Strength Functions
V. Werner et al... 117

NRF-Based NDA of Nuclear Material Using Monochromatic
γ-Ray Beam
T. Shizuma et al 127

Improving the Assay of ^{239}Pu in Spent and Melted Fuel Using
the Nuclear Resonance Fluorescence Integral Resonance
Transmission Method
C. T. Angell et al... 133

Laser Compton Scattering Gamma-Ray Beam Source
at NewSUBARU Storage Ring
S. Miyamoto et al... 143

Energy Calibration of Electron and Gamma-Ray Beams
at NewSUBARU-GACKO
T. Shima and H. Utsunomiya 151

A Paradigm for the Nondestructive Assay of Spent Fuel Assemblies
and Similar Large Objects, with Emphasis on the Role
of Photon-Based Techniques
A. M. Bolind 161

Medium Modification of α Cluster Size in ^{6}Li
T. Yamagata et al... 171

Noninvasive Reactor Imaging Using Cosmic-Ray Muons
H. Miyadera *et al* 177

Compton Radiation for Nuclear Waste Management and Transmutation
E. Bulyak and J. Urakawa 187

Compact Intense Neutron Generators Based on Inertial Electrostatic
Confinement of D-D Fusion Plasmas
K. Masuda *et al* 195

NRF Based Nondestructive Inspection System for SNM by Using
Laser-Compton-Backscattering Gamma-Rays
H. Ohgaki *et al* 203

Development of Measurement Methods for Detection of Special
Nuclear Materials Using D-D Pulsed Neutron Source
T. Misawa *et al* 209

SOFIA, a Next-Generation Facility for Fission Yields Measurements and
Fission Study. First Results and Perspectives
L. Audouin *et al* 217

Present Status of Nuclear Data for Nuclear Nonproliferation
M. Igashira 227

Development of the Experimental Photo-Nuclear Reaction Database in
Hokkaido University
A. Makinaga 235

Laser Compton Scattering Photon Beams and Other Gamma-Ray
Sources: Project for Coherent Gamma-Ray Source on Basis of
Femtosecond Laser at ILC MSU
V. G. Nedorezov and A. B. Savelév 243

Laser Driven Ion Acceleration Study in JAEA
K. Kondo 255

Status of New JENDL Photonuclear Data File
K. Kosako *et al* 261

Nuclear Research With Eγ≤15 MeV Photons
C. Rangacharyulu 269

IRIDE: Interdisciplinary Research Infrastructure Based on Dual
Electron Linac and Laser
M. Ferrario 279

Poster Presentations 289

A Study of the Nuclear Resonance Fluorescence Reaction Yield
Dependence on the Target Thickness of ^{208}PB
H. Negm *et al* 291

Photodisintegration Reactions with Linear Polarized γ-Ray Beam
T. Hayakawa *et al* 301

Test Experiment of γ-Ray Diffraction for Crystal Monochromators
S. Matsuba *et al* 309

Overview of Laser Compton-Scattered Photon Source at the cERL
R. Nagai *et al* 315

Development of a High-Brightness and High-Current Electron Gun
for High-Flux γ-Ray Generation
N. Nishimori *et al* 321

Design of ERL Spoke Cavity for Non-Destructive Assay Research
M. Sawamura *et al* 327

Measurements of Cosmic-Ray Muon-Capture X-Rays and Its
Application to Nuclear Material Detection
Y. Shimbara *et al* 333

Active Neutron-Based Interrogation System with D-D Neutron
Source for Detection of Special Nuclear Materials
Y. Takahashi *et al* 341

Closing summary 347

Summary Comments: Nuclear Physics and Gamma-ray
Sources for Nuclear Security and Nonproliferation
C. P. J. Barty 349

Program 351

List of Participants 357

Oral Presentations

Nuclear Science and Applications with the Next Generation of High-Power Lasers and Brilliant Low-Energy Gamma Beams at ELI-NP

S. Gales for the ELI-NP team

ELI-NP, "Horia Hulubei" National Institute for Physics and Nuclear Engineering, 30 Reactorului Street, RO-077125 Măgurele, jud. Ilfov, Romania

The development of high power lasers and the combination of such novel devices with accelerator technology has enlarged the science reach of many research fields, in particular High Energy, Nuclear and Astrophysics as well as societal applications in Material Science, Nuclear Energy and Medicine. The European Strategic Forum for Research Infrastructures (ESFRI) has selected a proposal based on these new premises called "ELI" for Extreme Light Infrastructure. ELI will be built as a network of three complementary pillars at the frontier of laser technologies. The ELI-NP pillar (NP for Nuclear Physics) is under construction near Bucharest (Romania) and will develop a scientific program using two 10 PW class lasers and a Back Compton Scattering High Brilliance and Intense Low Energy Gamma Beam, a marriage of Laser and Accelerator technology at the frontier of knowledge. In the present paper, the technical and scientific status of the project as well as the applications of the gamma source will be discussed.

Keywords: High power lasers; Compton back scattering low energy; high brilliance and narrow band width gamma beams; laser driven Nuclear Physics; NRF; photo-fission astrophysics and Nuclear applications.

1. Introduction

Ultra-intense laser fields with intensities reaching up to 10^{22-23} W/cm^2 are now able to produce typical high energy electrons, photons and ions currently in use in nuclear facilities, as demonstrated in laboratories across the globe. The emerging laser driven technologies are very promising in terms of cost, size and available parameter range. The European Strategic Forum for Research Infrastructures (ESFRI) has selected a proposal based on these new premises called "ELI" for Extreme Light Infrastructure. The construction of a large-scale laser-centered, distributed pan-European research infrastructure, involving beyond the state-of-the-art ultra-short and ultra-intense laser and gamma source technologies, was therefore funded for construction in September 2012. The three pillars of the ELI facility are built in Czech Republic, Hungary and Romania [1]. The first one, ELI-Beamlines is dedicated to the construction and characterization of secondary radiation sources, such as electrons, x-rays and protons, driven by high repetition rate ultra-intense lasers [2]. The second one is dedicated to the generation and use of attosecond electromagnetic pulses [3, 4]. The third pillar is ELI-Nuclear Physics (ELI-NP). Its mission covers scientific research involving two domains where only very few experimental results were reported until now: laser-driven experiments related to nuclear physics, strong-field quantum electrodynamics with associated vacuum effects and a Compton back-scattering high-brilliance and intense low-energy gamma beam, a marriage of laser and accelerator technology at the frontier of knowledge.

The ELI-NP research center will be located in Magurele, a town a few kilometers away from Bucharest, Romania. The facility, worth 295 million euros, will be developed in two phases and will be available for the users in 2018.

This paper presents the planned high-power laser system and the gamma-beam system parameters. It also describes the main scientific themes and associated experimental tools to be developed and implemented at the ELI-NP facility.

2. Technical description of the ELI-NP facility

ELI-NP hosts two major research infrastructure instruments: a high-power laser system (HPLS), with two arms – reaching up to 10 PW for each arm – and a gamma-beam system (GBS) that will provide very intense and narrow-band gamma-rays with energies up to 19 MeV. Eight experimental areas will be available for performing experiments, among which one experimental area will allow combined experiments. Figure 1 presents the block diagram of the facility.

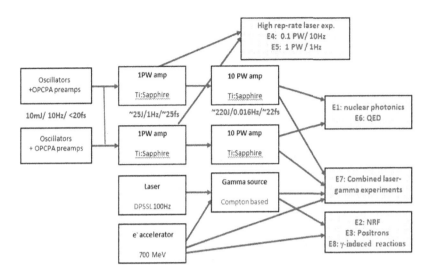

Fig. 1. Block diagram of the ELI-NP infrastructure.

2.1. High-Power Laser System (HPLS) at ELI-NP

The high-power laser system of ELI-NP will be constructed by an association between Thales Optronique SA France and Thales Romania. It consists of a chirped pulse amplification system at about 820 nm central wavelength, with a dual front-end architecture, in order to minimize down-time for the laser facility. Each of the identical front-ends can provide a high-contrast broadband laser pulse with energy in the tens of mJ range that is split and injected into the two parallel amplification chains. Each of the two parallel chains includes Ti: Sapphire amplifiers to bring the final output energy to the few hundreds of Joule level. Subsequently, the pulses are compressed to around a 22 fs pulse duration that implies a peak power of 10 PW for each of the two arms. The repetition rate for the two 10 PW outputs is one shot per minute.

Along the two amplification chains, additional outputs with corresponding optical compressors will be installed. Their corresponding power levels are 0.1 PW and 1 PW at repetition rates of 10 Hz and 1 Hz, respectively. The intensity in reach is 10^{23} W/cm^2, thus securing a properly defined laser pulse in both space and time for interaction with thin targets.

The overall control command of the HPLS will be TANGO-based. TANGO is a distributed control system, used for controlling any kind of hardware or software systems that is working based on the client-server model. The software is developed by the European Community of Synchrotrons: ESRF, SOLEIL,

ALBA, ELLETRA, etc. and is also used in other major projects like Laser Megajoule and Cilex–Apollon [5] from France.

The main focus of experiments involving HPLS as a source are:

Laser-driven Nuclear Physics experiments which will use accelerated heavy ions such as Th. The intended intensities in the focus are of the order of 10^{23} W/cm^2

Strong-Field QED physics: Electron behavior in ultra-intense laser fields will be studied. On one side, the tightly focused beams on solid targets will be used for electron positron pair creation studies. In this case, the requested specifications are similar to the ones for laser-driven Nuclear Physics experiments.

Combined laser-gamma experiments: here, the synchronization of the laser with the laser that drives the Compton backscattering process is a must. Moreover, coherent combination of the two laser pulses is requested on the long run, in order to provide the highest possible intensity on target.

Irradiated materials science experiments: laser pulses combined with various radiation types are requested, involving electrons, gamma, protons, neutrons and positrons.

2.2. *Gamma Beam System at ELI-NP (GBS)*

In designing a state-of-the-art gamma-ray beam source, several requirements have to be fulfilled, such as bandwidth as small as possible, strong increase of the peak brilliance and of the gamma beam flux, reduction of the gamma beam size. The gamma beam will be produced by the Compton backscattering of light photons on accelerated electrons. The photons are provided by an average power, high-repetition rate laser. The pulse energy is expected to be in the few hundreds of mJ range and frequency doubling of such laser pulses would provide green photons to be frequency up-shifted through the Compton backscattering process to 19 MeV. The pulse duration must stay above 1 picosecond, imposed by the narrow bandwidth condition for the backscattering gamma rays, namely, relative bandwidth smaller than 0.5%.

The electron accelerator will be a warm linac, with two acceleration stages of 360 MeV each. The energy of the Compton backscattered beam is given by the relation:

$$E_\gamma = 2\gamma_e^2 \cdot \frac{1 + \cos\theta_L}{1 + (\gamma_e\theta_\gamma)^2 + a_0^2 + \frac{4\gamma_e E_L}{mc^2}} \cdot E_L \tag{1}$$

Where

$\dfrac{4\gamma_e E_L}{mc^2} = $ recoil parameter;

$a_L = \dfrac{eE}{m\omega_L c} = $ normalized potential vector of the laser field;

$E = $ laser electric field strength; $E_L = \hbar\omega_L$

Thus in the case of a head-on collision with (θ_L=0) & backscattering (θ_γ=0), using a green laser with $E_L \sim 2.5$ eV (green) and an electron beam with $Ee \sim$ 300 MeV the gamma beam will have an energy of $E_\gamma <$ 3.5 MeV. The highest electron energies reachable will be of 720 MeV, allowing the production of up to 19 MeV gamma photons. A lower-energy gamma output (up to about 3.5 MeV) will be also available, with similar beam characteristics with the high-energy output.

Through a public tender procedure, the development of the gamma beam will be awarded to a company or consortium. The implementation will have three phases, with two intermediate deliveries before the final commissioning.

The minimum output specifications for the gamma beam (see Table 1) were established in a series of workshops and meetings organized with the scientific community interested in ELI-NP.

Table 1. Specified output parameters for the gamma beam system.

Type	Units	Range
Photon energy	MeV	$0.2 - 19.5$
Divergence	Rad	$\leq 2.0 \times 10^{-4}$
Average Bandwidth Gamma-Ray Beam	-	$\leq 5.0 \times 10^{-3}$
Time-Average Spectral-Density at Peak Energy	1/(s eV)	$\geq 5.0 \times 10^3$
Time-Average Brilliance at Peak Energy	1/(s mm^2 mrad^2 0.1% η,γ)	$\geq 1.0 \times 10^{11}$
Minimum Frequency Gamma-Ray Macropulses	Hz	≥ 100

3. Scientific program, experimental lay-out and instruments

A significant fraction of the international scientific community contributed to the shaping of the ELI-NP facility is a series of workshops [6]. The latest ones, held in June 2013, were centered on laser-driven experiments and on nuclear science with gamma beams and defined eight development directions for the facility. For each of them, writing of Technical Design Reports (TDRs) was triggered during the workshops.

In addition to fundamental themes, applications of HPLS and GBS are under study. Dosimetry, ionizing radiation metrology, radiation induced damage and gamma beams induced nuclear reactions are major active research area in nuclear physics and engineering. Their applications extend from the nuclear power plants to medicine and from space science to material science and to accelerators engineering.

The experimental areas will accommodate three types of experiments: laser-driven experiments will be performed at E1, E4, E5 and E6 experimental areas; gamma experiments will be performed at E2, E3 and E8 experimental areas, while combined experiments are planned at E7 experimental area. A layout of the experimental areas is displayed in Figure 2.

3.1. *Laser-driven nuclear physics experiments*

As a leading facility in laser-driven nuclear physics, ELI-NP will take advantage of the specific properties of laser driven radiation production, such as ultra-short time scale and the relatively broadband spectrum of secondary radiation at the experimental areas E4 (two 100 TW pulses at 10 Hz) and E5 (two 1 PW pulses at 1 Hz). Further specificity of the proposed experimental environment at ELI-NP is to simultaneously provide two or more types of radiation on the same target (e.g. laser-accelerated electrons and laser-accelerated ions).

Ion driven nuclear physics research will be performed at the E1 experimental area. Here, laser-accelerated heavy ions will drive nuclear reactions and further experiments related to astrophysics. The flagship experiment involves production of neutron rich isotopes using fission of Thorium and subsequent fusion process, to shed light on one of the most important questions in today's physics: the formation of heavy elements (beyond Fe) in the universe. The needed data for the formation of heavy elements in the region of lead (Z=82) and beyond are masses and lifetimes of key isotopes, e.g. very neutron-rich isotopes (15 neutrons away from the stability line), and the fusion cross section even with secondary beams of radioisotopes is so low that in practice these key measurements are not within reach today.

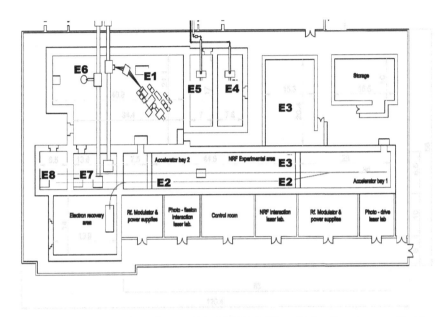

Fig. 2. Layout of the experimental areas of the ELI-NP facility, with the dimensions specified in meters

Using HPLS with a CD_2 production target and a very thin thorium reaction target, one may reach fusion of two very neutron-rich isotopes originating from the fission of Th (for example fusion of Z= 35) leading to this unknown mass region [7]. Such experiments require significant experimental development in the field of laser-driven ion acceleration according to new acceleration mechanisms in order to produce intense heavy-ion bunches in the 5–10 MeV/n energy range relevant for fission and fusion reactions.

The experimental area E6 will host experiments related to strong-field quantum electrodynamics. High intensity on solid targets, up to 10^{23} W/cm^2 is intended. The type of experiments will imply a different set of diagnostics tools, compared to the ones at E1 where ion-driven nuclear physics experiments will be performed. Here, electron-positron production in laser-irradiated solid targets and their subsequent behavior will be the main object of study [8].

The combined experiments from E7 area will host a vacuum chamber where one pulse from each 10 PW HPLS output will be focused. The two pulses will be synchronous, allowing both laser-laser experiments but also laser-gamma. The envisaged experiments are related to studies of the vacuum in strong fields, such as studies of vacuum birefringence, gamma-assisted pair creation in strong laser field and dark matter studies [9].

3.2. Nuclear Science and applications with high brilliance low-energy gamma beams

The ELI-NP –gamma beam experimental program will have allow us to explore new territory in the field of nuclear resonance fluorescence (NRF) and experiments above the particle separation threshold, such as studies of giant resonances, nuclear astrophysics reactions, and photo-fission experiments. A schematic lay-out of such experiments is displayed in Figure 3. The incoming narrow band width beam excites a single excited state, whose decay is studied in the experiment. The excited state can be below or above the particle separation energy, which lies at about 8 MeV. In the former case, the NRF method is applied and, in the latter case, induced reactions, such as (γ, n), (γ, p),

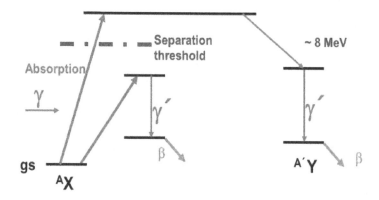

Fig. 3. Principle of NRF experiments.

During the last workshop held in Bucharest-Magurele in June 2013 (see Ref. 6) the main lines of this future scientific program were discussed, working groups were formed towards completion of TDRs based on the following science issues.

The low-energy gamma experiment area E2 (E_γ<3.5 MeV) shall be mainly dedicated to **Nuclear Resonance Fluorescence (NRF) experiments and applications**. Due to the brilliance of the gamma-ray beam, significantly increased with respect to existing facilities, experiments on materials whose availability is very limited will become feasible. For example the photo-response below the particle separation energy is currently investigated in NRF experiments [10] at existing gamma-beam facilities, such as bremsstrahlung facilities at the S-DALINAC electron accelerator in Darmstadt (see *e.g.* Ref. [11]) or at the High Intensity Gamma-ray Source (HIGS) at the TUNL facility at Duke University (see *e.g.* Ref. [12]). The advances in gamma-ray beam brilliance at ELI will increase the sensitivity of NRF experiments and thus it offers the opportunity to perform NRF studies on small target samples. This opens up an entire new area of applicability of the NRF method to materials that may be available only in quantities of a few mg, like long-lived radioactive isotopes of heavy actinides.

The high brilliance, small bandwidth and tunable energy of the low-energy gamma beam may revolutionize the characterization of nuclear materials. Figure 4 illustrates the principle of this method.

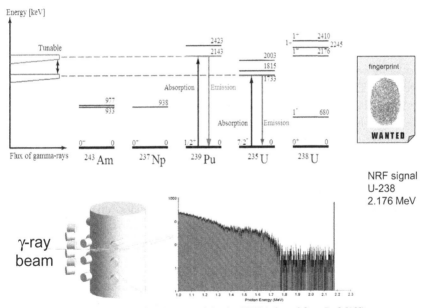

Fig. 4. Fingerprint of nuclear material using NRF (adapted from Ref. [13]).

The tunable-energy gamma-ray beam can be locked on the known excitation energy of the isotope to be located and identified (each isotope has a unique finger print) and the energy response of the detectors will clearly indicate the location and presence or absence of the isotope in the bulk material (*e.g.* in a nuclear waste container). This technique will be extremely useful in the management of sensitive Nuclear Materials and radioactive waste for isotope-specific identification of $^{238}U/^{235}U$ and ^{239}Pu. It will allow the scan of containers for nuclear material and explosives, the inspection of spent fuel elements, and the quantitative measure of the final ^{235}U, ^{238}U content in order to optimize the geometry towards the longer use (20%) of fuel elements in the reactor core.

The E8 experiment area at ELI-NP is dedicated to experiments that can use the entire range of gamma ray energies, but mainly the higher energies 3–19 MeV. An intense and high-energy resolving γ–ray beam from ELI-NP will open up new horizons for the **investigation of the nuclear photo-response at and above the separation threshold**. An example for such studies is the detailed investigation of the pygmy dipole resonance (PDR) above and below the particle threshold, which is very essential for nucleosynthesis in astrophysics. The PDR occurs close to the neutron-emission threshold and its decay is governed by the coupling to the large number of states around the threshold. Both the GDR and the PDR can be covered within the energy range of the ELI-NP beams. In the experiments the excitation functions for elastic and inelastic scattering will be measured, revealing possible fine-structures/splitting of PDR and GDR. The excitation function with high resolution for (γ,n) and (γ,charged-particle) channels, allows one to determine the branching ratios for various decay channels. The polarized beam will also allow determining the $E1$ or $M1$ type of excitation for the observed structures [1, 6].

Photo-fission is a topic where ELI-NP gamma beam will bring significant advances. So far bremsstrahlung was used to induce fission of actinide nuclei. Two classes of experiments have been identified:

a) High-resolution photo-fission studies in the actinides, investigation of the second and third potential minima, angular and mass distribution, cross-sections, studies of rare photo-fission events, such as triple fission, highly asymmetric fission, *etc.*

b) Low-energy gamma beams are fully efficient at 15 MeV for producing short-lived and refractory elements in thin U targets. using a gas-cell catcher (IGISOL technique [14]) with high efficiency, due to the low ionizing power of pure γ-ray beams. Limited investments, minimizing radioactivity: a real niche! After their separation, the nuclei of interest will be transported to different measurement stations.

Laboratory astrophysics experiments aiming at explaining the nucleosynthesis processes will be possible, through direct or inverse reactions.

Reactions relevant for the p- and r-processes will be investigated, to advance the explanation of the formation of a large part of the known elements in the Universe. All p-nuclei can be synthesized from the destruction of pre-existing nuclei of the s- and r-type by a combination of (p, γ) captures and (γ,n), (γ,p) or (γ, a) photo-reactions [15]. In particular, charged-particle detector systems, needed to measure nuclear reaction cross-sections of the proton and alpha burning processes and – most importantly – the $^{12}C(\alpha,\gamma)$ reaction cross-section relevant for stellar helium burning, are being investigated [16].

In order to carry out the scientific program discussed above, a number of different state-of-the-art instruments are being considered. These include: a high-resolution spectrometer of (segmented) large HPGe (clover) detectors, combined with good timing *e.g.* LaBr3 detectors, a spectrometer with medium resolution of large LaBr3 detectors and a neutron detector array, a tape station and a close-geometry spectrometer for high-resolution decay studies, a 4π charged-particle array of segmented DSSSD detectors and a TPC gas cell for astrophysics reaction measurements.

In addition, **a variety of applied research proposals** have been received by ELI-NP using low-energy brilliant intense gamma, neutron and positron beams, which will open new fields in materials science and life sciences.

ELI-NP will feature a brilliant positron beam, the E3 experiment area being dedicated to research based on this beam, including imaging and applications [1,6].

The new production schemes of medical isotopes (*e.g.*, ^{99}Mo currently used in therapies, ^{195m}Pt for nuclear imaging to determine efficiency of chemotherapy, and ^{117m}Sn, an emitter of low-energy Auger electrons for tumor therapy) via (γ, n) processes may also reach socio-economical relevance.

Computerized tomography with gamma–ray beams for non-destructive inspection of objects, will benefit also from high-energy quasi–monochromatic and high beam intensity to shorten the scanning time

4. Conclusions

The ELI-NP facility mixes two research facilities with parameters beyond the state of the art, namely a high-power laser system with two amplification arms to deliver 10 PW and intensities on the target in the range of 10^{23} W/cm^2 at least every minute, and a gamma beam system to deliver up to 19 MeV photons. Their outstanding performances will allow approaching a virgin science field, at the frontier between the strong-field QED and nuclear physics.

Benefiting from the support of a large number of specialists across the globe, the ELI-NP facility is on track with Technical Design Reports and construction of the experimental areas. Commissioning is expected to take place in 2018.

Acknowledgments

This work is supported by the Extreme Light Infrastructure – Nuclear Physics (ELI-NP) – Phase I, a project co-financed by the Romanian Government and European Union through the European Regional Development Fund.

References

1. Gérard A. Mourou, Georg Korn, Wolfgang Sandner, John L. Collier (eds.) *ELI – Extreme Light Infrastructure: Science and Technology with Ultra-Intense Lasers Whitebook*, (THOSS Media GmbH, 2011).
2. B. Rus, P. Bakule, D. Kramer, G. Korn *et al.*, 87801T (May 7, 2013), doi:10.1117/12.2021264.
3. S. Banerjee, M. Baudisch, J. Biegert, A. Borot, A. Borzsonyi et al., *CLEO: 2013, OSA Technical Digest* **CTu2D.6** (Optical Society of America) 2013.
4. C. L. Arnold, F. Brizuela, A. Borot, F. Calegari, *CLEO: 2013, OSA Technical Digest* **JTh2A.13** (Optical Society of America) 2013.
5. G. Cheriaux, F. Giambruno, A. Freneaux, F. Leconte,*et al*, Apollon-10P: status and implementation, in: *AIP Conference Proceedings on Light at Extreme Intensities* **1462**, 78 (2012).
6. http://www.eli-np.ro/documents/
7. P.G. Thirolf *et al.*, *EPJ Web of Conferences*, **38**, 08001
8. E. Esarey, C. B. Schroeder, and W. P. Leemans, *Rev. Mod. Phys.* **81**, 122
9. A. Di Piazza, C. Müller, K. Z. Hatsagortsyan, and C. H. Keitel, *Rev. Mod. Phys.* **84**, 1177
10. U. Kneissl, N. Pietralla, A. Zilges, *J. Phys. G* , R217 32 (2006)
11. D. Savran *et al.*, *Phys. Rev. Lett.* **100**, 232501 (2008).
12. A. P. Tonchev *et al.*, *Phys. Rev. Lett.* **104**, 072501 (2010).
13. R. Hajima *et al.* , *J. Nucl. Sci. Tech.* **45**, 441 (2008).
14. J. Aystö *et al.*, Phys. Rev. Lett. **69**, 1167 (1992).
15. K.-L. Kratz *et al.*, *Astrophys. J.* **403**, 216 (1993).
16. W. A. Fowler, *Rev. Mod. Phys.* **56**,149 (1984).

New Non-Intrusive Inspection Technologies for Nuclear Security and Nonproliferation

Dr. Robert J. Ledoux
Passport Systems, Inc.
Billerica, MA 01862, USA
ledoux@passportsystems.com
www.passportsystems.com

Comprehensive monitoring of the supply chain for nuclear materials has historically been hampered by non-intrusive inspection systems that have such large false alarm rates that they are impractical in the flow of commerce. Passport Systems, Inc. (Passport) has developed an active interrogation system which detects fissionable material, high Z material, and other contraband in land, sea and air cargo. Passport's design utilizes several detection modalities including high resolution imaging, passive radiation detection, effective-Z (EZ-3D™) anomaly detection, Prompt Neutrons from Photofission (PNPF), and Nuclear Resonance Fluorescence (NRF) isotopic identification. These technologies combine to: detect fissionable, high-Z, radioactive and contraband materials, differentiate fissionable materials from high-Z shielding materials, and isotopically identify actinides, Special Nuclear Materials (SNM), and other contraband (e.g. explosives, drugs, nerve agents). Passport's system generates a 3-D image of the scanned object which contains information such as effective-Z and density, as well as a 2-D image and isotopic and fissionable information for regions of interest.

1. New Non-intrusive Active Inspection Technologies for Nuclear Security and Nonproliferation

1.1. *Overview*

An important enabling technology for Passport's active non-intrusive active inspection technologies is the use of a high intensity, high duty cycle electron source which is magnetically scanned across a collimated radiator. A high duty cycle accelerator allows for the detection of individual photons and neutrons with very high average currents. Thus, detector pile up is avoided even at average beam intensities which are greater than 100 times those currently available from pulsed LINACs. A schematic of the Passport SmartScan™ scanner is shown in Figure 1. It employs an IBA TT100 electron accelerator typically producing variable current (up to 4 mA) 9 MeV electron beams. The detection sub-systems include a transmission X-ray imager, a fixed collimated array (both sides of cargo) of high-energy photon detectors, fixed array of high-energy neutron detectors, and a movable array of high energy resolution germanium detectors.

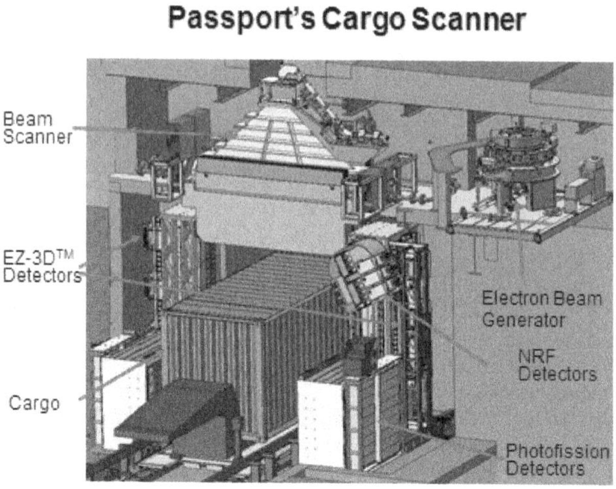

Passport's Cargo Scanner

Figure 1: Schematic of Passport's SmartScan™ System for cargo inspection

A summary of these new inspection modalities is given in Table 1. The first column indicates if the modality is used in the initial (primary) or prolonged (secondary) screening. Table 1 also indicates what types of algorithms or methodologies are used to transform the listed input data type into various materials properties represented as three dimensional (3-D), two dimensional (2-D) or isotopic composition of selection region of interest (ROI). These modalities can be used individually or collectively to extract a quantitative measure of anomaly classification. Examples of data fusion will be given later in this document.

It is worth noting that the sensors utilized in these advanced non-intrusive inspection modalities are all commercially available as mature detector technologies. As stated above, the use of a high duty-cycle, high intensity electron accelerator (IBA TT100) enables the measurement of the energy and angular distributions of the backscattered photons and prompt neutrons. These measurements combined with the algorithms listed in Table 1 allow for material discrimination capabilities via EZ-3D™ and PNPF and the materials identification of NRF. Details of these new technologies are given inreferences: W. Bertozzi, S. E. Korbly, R. J. Ledoux, and W. Park, *Nuclear resonance fluorescence and effective Z determination applied to detection and imaging of special nuclear material, explosives, toxic substances and contraband*, Nuclear Instruments and Methods B 261, 331 (2007)

W. Bertozzi, W. Franklin, S. E. Korbly, R. J. Ledoux, R. Niyazov and D. R. Swenson, *Accelerators for Homeland Security*, International Journal of Modern Physics A 26, Nos. 10 & 11, 1713 (2011).

Table 1: Overview of Advanced Non-Intrusive Screening Technologies

Scan	Algorithm	Input	Functionality / Output
Initial	EZ-3D™ Reconstruction	Medium-resolution energy spectrum	3-D density and Effective-Z map Anomaly identification/3D location
Initial	Transmission X-ray	Medium/High spatial resolution transmission image	Anomaly 2D location & density Shape/edge recognition
Initial	Portal Networked Detection System	Medium-resolution passive spectrum	Identification and localization of radioactive sources
Initial & Prolonged	Photofission	High-resolution energy spectrum	Identifies presence of fissionable material
Prolonged	NRF 3D	Digitized pulses from liquid organic scintillator	Complete isotopic composition in the region-of-interest
	Anomaly Classification	Output of NRF 3D, PNPF, EZ-3D™ and transmission algorithms	Performs data fusion, classifies anomaly as threat or innocuous, predicts detect/clear time

The method for generating the 3-D atomic number and density map (EZ-3D™) and the 2-D map of fissionable materials (PNPF) is illustrated in Figure 2. The 9 MeV electron beam is magnetically scanned across a radiator at fixed highly collimated ports. While the electron beam dwells on each port, a fixed array of collimated high-energy photon detectors views the complete path of the "pencil" beam through the cargo. Thus, at each beam port, the energy distribution of backscattered photons is viewed from the voxels formed by the intersection of the detectors field of view and the photon beam. One full sweep of the beam across the container cross section yields a measurement of a slice of the cargo. The collection of slices forms a 3-D set of backscattered photons from 10's of thousands of voxels. This data set is reconstructed to obtain a 3-D map of the atomic number and density of the inspected volume. Further, at each beam position, the energy distribution of high-energy neutrons is obtained. Through an analysis of the energy distribution of these high energy neutrons, a unique signature of fissionable material is obtained in 2-D.

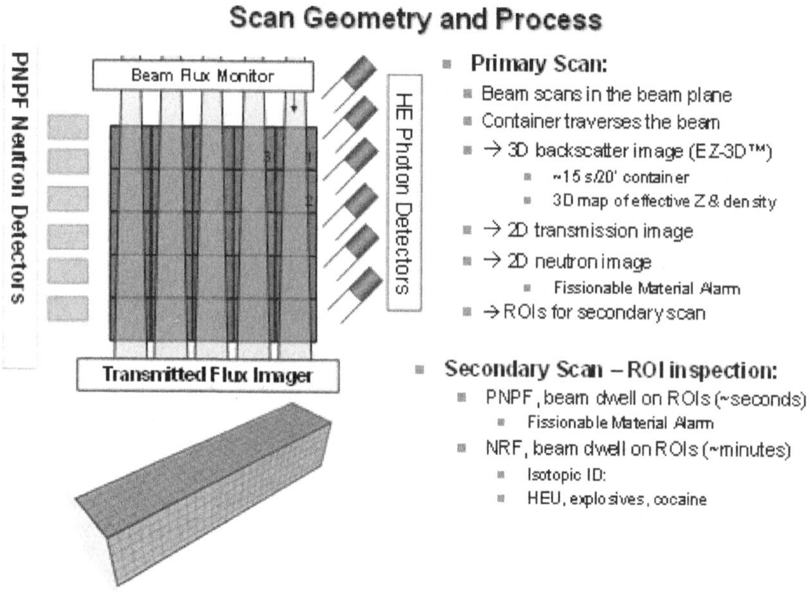

18

Figure 2: Schematic representation of the magnetic beam scanner and detector placement

1.2. Non-Intrusive Inspection Example: Initial Scan

Shown in Figure 3 is an example of an initial scan of a truck loaded with various bulk cargoes with threat and contraband articles hidden inside. The scan time was approximately 30 seconds. The translucent color of each voxel is a measure of its average atomic number with color index shown at the left of the truck. The opacity of each region is a measure of the average density. ROI are completely opaque with an anomaly color code to the right of the truck. These anomalous regions were determined based on a database of pre-determined regions of Z and density ([Z, ρ]) and/or anomalous regions of Z and density as compared to surrounding cargo. The user can generate custom regions of [Z, ρ] that generate automated flagging of ROIs. The anomalies chosen in this example represent nuclear threats (depleted uranium), possible high-Z gamma shielding, explosives, illegal contraband (cocaine simulant) and economic contraband (non-declared tobacco).

The intrinsic 3-D nature of EZ-3D™ allows for much greater sensitivity to [Z, ρ] as compared to dual-energy 2-D radiography since the voxel by voxel signal is not "diluted" by averaging over projection. Moreover, the exact 3-D placement of the anomaly provides both contextual information regarding its location within the cargo and the exact position information needed to remove the object from the cargo.

EZ-3D Volumetric Data of Density and Effective Z

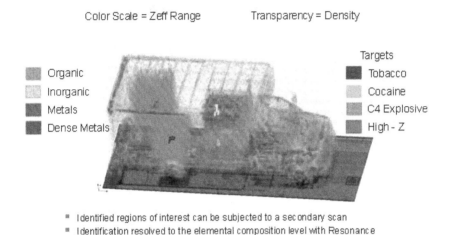

Color Scale = Zeff Range   Transparency = Density

Organic   Targets
Inorganic   Tobacco
Metals   Cocaine
Dense Metals   C4 Explosive
                                  High - Z

* Identified regions of interest can be subjected to a secondary scan
* Identification resolved to the elemental composition level with Resonance Fluorescence

Figure 3: Initial Scan of truck with various cargos and automatically detected anomalies

Shown in Figure 4 is a projection of the full 3-D effective Z and density data and detected high-Z anomalies onto the transmission X-ray image. This allows for a detailed comparison of the two measurements. The EZ-3D™ data can be sliced along any plane through the volume. Alternately, ROI can be analyzed in greater detail in 3-D through the removal from the display of outer layers of the cargo surrounding the anomaly and viewing of the anomaly from any perspective.

2D Transmission Views with High Z and EZ-3D Overlay

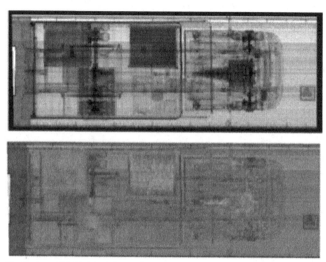

Figure 4: Projection of EZ-3D™ information onto 2-D transmission image

As part of the initial scan, the PNPF 2-D measurements are also obtained. An example of the projection of both EZ-3D™ and PNPF data over the transmission X-ray image is displayed in Figure 5. There are many regions of high projected density, however, only two regions indicate the presence of high-Z material (colored regions in upper part of figure) via the EZ-3D™ data. Surprisingly, the PNPF projection (lower overlay) indicates that the high-Z anomaly with the lowest projected density is actually fissionable material! The transmission image under the fissionable material threat does not show an unusually high density in projection since it is a very thin piece of depleted uranium (DU). The other high-Z anomaly is the lead in the truck battery. Therefore, the fusion of the EZ-3D™ and PNPF measurements automatically determine the following: 1) there is fissionable material in the cargo, 2) it is very thin and 3) it does not have high-Z (gamma ray) or large quantity of hydrogenous material (neutron) shielding material surrounding it.

Automated High Z Material Detection

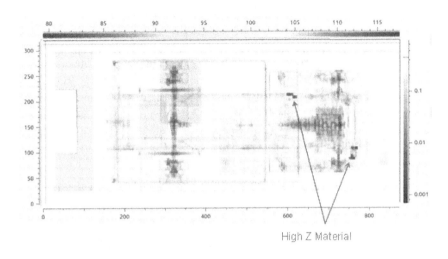

High Z Material

Automated Fissionable Material Detection

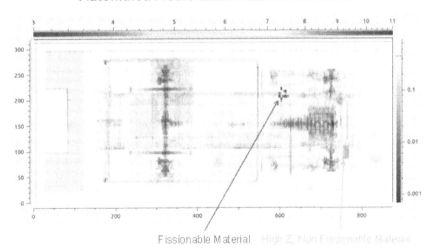

Fissionable Material   High Z, Non Fissionable Material

Figure 5: Upper: Projection of EZ-3D™ high-Z regions on transmission X-ray image, Lower: same for PNPF data with clear identification of fissionable material at one high-Z location

1.3. Non-Intrusive Inspection Example: Secondary Scan

A secondary scan may be performed for any anomalies automatically determined from the initial scan, or manually selected by the operator. The 3-D coordinates are extracted from the EZ-3D™ image. This region of interest is then automatically positioned under the beam scanner, the appropriate beam port selected and the collimated NRF array is positioned to view the ROI. Two examples of NRF interrogation of regions of interest are shown in Figure 6 & Figure 7. For each Figure, the EZ-3D™ ROI is shown on the left and a fidelity plot is shown on the right for representative materials. A positive fidelity (fidelity of 2 is 95% confidence) indicates the presence of a material, negative values the absence of the material.

In Figure 6, EZ-3D™ has identified two ROI that have a density close to 1 g/cc and an average Z less than 10. What are these objects? The NRF fidelity plots of each object clearly and automatically identify one object as being water and the other as alcohol (it was 80 proof vodka). These measurements were each obtained in approximately 1 minute. The representative materials are contained in a user specified library which contains both elemental materials (O, N, Pb, etc.) as well as compounds. The signatures of important compounds are directly measured, i.e., if you want the NRF signature of a banana, stick the banana in the beam, measure and store! New compounds can be easily added to the library as well as their associated alarm level. Figure 7, the ROI of interest has been flagged since it consists of a low-Z material with density greater than 1.3 g/cc embedded in a low-Z cargo that has a much lower average density. NRF interrogation of this anomaly reveals that it is consistent in composition to C4 explosives.

It is important to note that EZ-3D™, PNPF and NRF are spectroscopic measurements. Through data fusion, the bulk attenuation for photons and limits for high-energy neutron attenuation can be determined. The source terms have been measured for PNPF and NRF and the photon beam output is known. These data can be used to determine the expected time to detect or clear a particular mass of a previously measured material. This provides a quantitative measure by which to clear cargo for the presence of materials of interest which is not available in conventional X-ray inspection.

Secondary Scan Contraband Identification Example

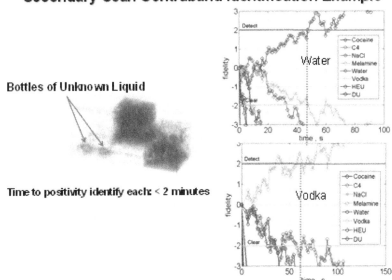

Figure 6: Materials identification for two liquids identified in the initial scan

Secondary Scan Explosive Identification Example

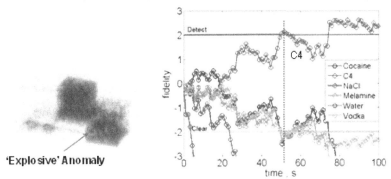

- Potential explosive anomaly detected by EZ-3D
  - Density and Effective Z by voxel
- Positively identified as C4 in < 1 minute with RF

Figure 7: Materials identification of an explosive anomaly

24

1.4. Summary

The SmartScan™ combines multiple non-intrusive inspection technologies to obtain the following advantages over existing scanner:
- Quickly clears most cargo in the initial scan
- Automatically identifies suspect threat items / regions of interest
- Automatic or directed secondary screening - elemental analysis
- Provides a very high level of confidence that whenthe cargo is opened the targeted item will be quickly found – Both material and 3-D location
- Gives a high level of confidence that when a scanned cargo leaves it is truly "cleared" - no targeted materials are inside
- Even more effective when combined with intelligence and targeting

An artist's rendition of a SmartScan™ facility being constructed at the Port of Boston is shown in Figure 8.

Figure 8: SmartScan™ pilot facility at Boston's container port

Acknowledgments

The author would like to thank his many colleagues who contributed to the results discussed here:

William Bertozzi,James B. Costales, Areg Danagoulian, Wilbur Franklin, Krzysztof Kamieniecki, Stephen E. Korbly, Rustam Niyazov, Dave R. Swenson, &Cody M. Wilson.

All statements in this paper are solely those of the author.

Development of Gamma-Ray Nondestructive Detection and Assay Systems for Nuclear Safeguards and Security at JAEA

Ryoichi Hajima

Quantum Beam Science Directorate, Japan Atomic Energy Agency
Tokai, Ibaraki 3191195, Japan
hajima.ryoichi@jaea.go.jp

Nondestructive detection and assay of nuclide is one of the promising applications of energy-tunable gamma-rays from laser Compton scattering. In JAEA, we are developing technologies relevant to the gamma-ray non-destructive assay, which include a high-brightness gamma-ray source based on advanced laser and accelerator technologies and gamma-ray measurement techniques optimized for highly radioactive samples. In this paper, the status of the above R&D's is reviewed.

Keywords: Gamma-ray; laser Compton scattering; nuclear resonance fluorescence; nuclear safeguards; nuclear security.

1. Introduction

Generation of mono-energetic and energy-tunable gamma-ray beams from laser Compton scattering (LCS) is becoming a common technology thanks to progress in modern electron accelerators and laser systems. The LCS gamma-ray is distinct from other conventional gamma-ray sources such as bremsstrahlung and radio isotopes in its energy tunability, narrow energy width and small divergence[1].

We are proposing applications of LCS gamma-ray beams to nondestructive measurements of nuclear materials. Utilizing this LCS gamma-ray beam in combination with nuclear resonant fluorescence (NRF), we can make nondestructive measurements of arbitrary nuclides[2]. Isotope-specific NDA systems based on LCS-NRF can be realized in several ways: resonance scattering, resonance transmission, integral resonance transmission and photo-fission[3]. A high-flux and energy-tunable gamma-ray source is a key technology for all the NDA systems.

For the generation of high-flux gamma-rays via laser Compton scattering, collision density of laser and electron beams must be increased. Therefore, a small emittance and high-current electron beam and a high-power laser are inevitable. Since average flux is essential rather than peak flux in the NDA applications, the electron beam current and the laser power should be evaluated in a sense of average values.

An energy-recovery linac (ERL) is the most suitable apparatus to accelerate electron beams of small emittance and high-average current[4]. The ERL is able to generate a high-quality electron beam with a high-average current. An electron beam from an injector is accelerated by a time-varying radio-frequency (RF) field stored in a superconducting linear accelerator and subsequently is transported to a recirculation loop. After the recirculation, the electron beam is injected again to the superconducting accelerator with the deceleration RF phase. The recirculated electrons are decelerated and feed back the energy to the superconducting RF cavity. This recycled RF energy is again used to accelerate subsequent electrons. The ERL is thus composed of an injector, a superconducting linac and an energy recovery loop. Fig. 1 shows a schematic view of the LCS gamma-ray source designed on the basis of the ERL.

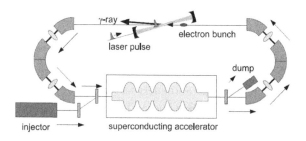

Fig. 1.   Laser Compton scattered gamma-ray source based on an energy-recovery linac

In order to obtain a high-flux gamma-ray beam, we employ a laser enhancement cavity for laser Compton scattering. The laser enhancement cavity is a high-finess Fabry-Perot optical cavity to store a high-power laser. Laser photons stored in the cavity interact with electrons many times to generate gamma-rays.

For nondestructive measurements of Pu in spent nuclear fuel, we design a gamma-ray source to produce a gamma-ray beam of $\sim$2 MeV at a flux of 1×10^{13} ph/s. The gamma-ray source consists of a 350-MeV 10-mA ERL and a laser enhancement cavity storing a laser beam of 700-kW average power[5].

2. R&D status for LCS-gamma ray NDA

In order to obtain a high-flux and high-intensity gamma-rays from a LCS source based on an ERL, there are key technologies to be developed: generation and acceleration of small-emittance and high-average current electron beams in an ERL, storage of high-power laser pulses in a laser enhancement cavity, collision of electrons and laser photons at a small spot size, stabilization of beams, and so on. The followings are the development status of these key technologies.

2.1. *Electron gun and superconducting accelerator*

An electron gun optimized for the ERL has been designed and fabricated at JAEA. which is a photocathode DC electron gun to generate small-emittance electron beams at high-average current. We have demonstrated an operation of the electron gun at the world-highest voltage, 500 kV, and a high-average current up to 10 mA[6]. After the successful demonstration of beam generation, the gun was shipped to High Energy Accelerator Research Organization (KEK) and installed at Compact ERL (cERL), a test accelerator at KEK. We started a beam operation of cERL from April, 2013. Now, the gun provides electron beams for the daily operation. Figure 2 shows the electron gun installed at cERL.

A superconducting accelerator (SCA) for high-average current electron beams is another key component in an ERL. We have a collaborative efforts with KEK for the superconducting accelerators. So far, two types of superconducting accelerators have been developed for the ERL, one for the injector and the other for the main linac, both of them are operated at 1.3 GHz RF.

The injector SCA is a 2-cell 3-cavity type to capture the 500-keV electron beam from the gun and accelerate it to ~5 MeV. The amplitude and phase of RF are adjustable individually for the best acceleration of the beam. Commissioning of the injector SCA started from May, 2013. We have confirmed beam acceleration to 5.7 MeV at the exit of the injector SCA[7]. Figure 3 shows the injector SCA.

The main linac SCA, a 9-cell 2-cavity type, was assembled from August to October 2012, and installed at cERL as shown in Fig. 4. A high-power test was conducted to demonstrate accelerating voltage of 14.2 MV and 13.5 MV in each of the two cavity[8].

Commissioning of cERL has been conducted and we have achieved an electron beam transportation through the injector and the recirculation loop as designed[9]. During the commissioning, we have checked all the accelerator components and fixed minor issues such as electron orbit distortion at the injector by stray magnetic fields from cold cathode vacuum gauges. Now, cERL is operated regularly and we are continuing studies to improve the beam performance.

Fig. 2. The 500-kV DC photocathode electron gun installed at Compact ERL

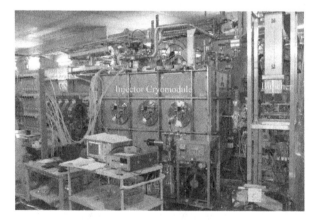

Fig. 3. The superconducting RF accelerator installed at cERL injector

Fig. 4.   The superconducting RF main linac installed at cERL

2.2. *High-Power Laser*

A high-power mode-locked laser for the laser Compton scattering is also under development at JAEA Kansai. The laser consists of a mode-locked oscillator and 4-stage amplifiers, all of which utilize Yb-doped fiber as laser gain media. We have completed the oscillator and 4-stage amplifiers to produce laser pulses with the following properties: wavelength 1030 nm, average power over 100W, repetition 80 MHz[10]. Figure 5 shows the laser system.

2.3. *Demo-experiment planned at Compact ERL*

We plan to make a demo-experiment of high-flux LCS photon generation at cERL. We will install a laser enhancement cavity at the recirculation loop of cERL to generate high-energy photons via laser Compton scattering. The LCS photons are transported to an experimental room for evaluation of the LCS beam. Figure 6 shows a floor layout of the cERL for the LCS experiment.

The electron gun and superconducting accelerators have been installed at the cERL as described above. We plan to install the laser and laser enhancement cavity in 2014. The LCS experiment is schedule in March 2015. Since the designed electron energy is 35 MeV, LCS photons will have an energy of 22 keV. However, we consider it is a demonstration of high-flux laser Compton light sources for future gamma-ray applications of nuclear material NDA systems. Increasing the LCS photon energy is simply

Fig. 5.   Mode-locked fiber laser developed for the LCS experiment

achievable by adding superconducting accelerators for the higher electron energy.

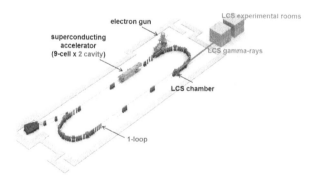

Fig. 6.   Schematic view of Compact ERL and apparatus for the LCS experiment

3. Summary

We have proposed a non-destructive detection and assay of nuclear material based on nuclear resonance fluorescence triggered by LCS gamma-ray beams. This technique can be applied to non-destructive detection and assay of fissile materials and other isotopes for nuclear security and nuclear safeguards purposes. We are developing technologies relevant to the LCS gamma-ray generation such as energy-recovery linac and high-power mode-locked laser. We plan to generate a LCS photon beam at Compact ERL as a demo-experiment of high-flux photon generation from a LCS gamma-ray source based on an ERL.

The author gratefully acknowledge T. Hayakawa, T. Shizuma, C.T. Angell, M. Sawamura, R. Nagai, N. Nishimori, S. Matsuba, N. Kikuzawa, M. Mori, A. Kosuge, M. Seya at JAEA for the development of gamma-ray NDA systems, Prof. H. Kawata, Prof. Y. Kobayashi and Prof. J. Urakawa at KEK for their collaboration on the LCS experiment at cERL. This work is supported in part by Ministry of Education, Culture, Sports, Science and Technology (MEXT).

References

1. G.A. Krafft and G. Priebe, Rev. Accl. Sci. Tech. **03**, 147 (2010).
2. R. Hajima, J. Nucl. Sci. Tech. **45**, 441 (2008).
3. R. Hajima et al., Eur. Phys. J. Special Topics, **223**, 1229 (2014).
4. R. Hajima, Rev. Accl. Sci. Tech., **03**, 121 (2010).
5. R. Hajima, Nucl. Instr. Meth. A **608**, S57 (2009).
6. N. Nishimori et al., Apl. Phys. Lett. **102**, 234103 (2013).
7. E. Kako et al., Proc. IPAC-2013, 2140 (2013).
8. K. Umemori et al., Proc. of SRF-2013, 855 (2013).
9. S. Sakanaka et al., Proc. of ERL-2013, 16 (2013).
10. M. Mori et al., Proc. CLEO-PR-2013 and OECC-2013, paper:MD1-4 (2013).

Photonuclear Reaction Studies at HIγS: Developing the Science of Remote Detection of Nuclear Materials

C. R. Howell

Department of Physics, Duke University and Triangle Universities Nuclear Laboratory,
Durham, NC 27708, USA
howell@tunl.duke.edu

Development of gamma-ray beam interrogation technologies for remote detection of special nuclear materials and isotope analysis requires comprehensive databases of nuclear structure information and gamma-ray induced nuclear reaction observables. Relevant nuclear structure details include the energy, spin and parity of excited states that have significant probability for electromagnetic transition from the ground state, i.e, the angular momentum transferred in the reaction is $\Delta I \leq 2$. This talk will report recent Nuclear Resonance Fluorescence (NRF) measurements to identify and characterize new low-spin states in actinide nuclei at energies from 1 to 4 MeV, which is the energy range most important for remote analysis methods. These measurements are carried out using the nearly mono-energetic linearly polarized gamma-ray beam at the High Intensity Gamma-ray Source (HIγS) at the Triangle Universities Nuclear Laboratory. Also, studies of the (γ, n) reaction on a variety of nuclei with linearly polarized beams at HIγS indicate that this reaction might be used to discern between fissile and non-fissile materials. This work will be described. In addition, an overview will be given of a concept for a next generation laser Compton-backing scattering gamma-ray source to be implemented as an upgrade to increase the beam intensity at HIγS by more than an order of magnitude.

Keywords: Photonuclear; photofission; fission; actinide; nuclear resonance fluorescence.

1. Introduction

Systems for remote assaying of objects will require applications of both passive and active integration techniques for identification of isotopes and accurate determination of the relative isotopic composition of materials. Assaying approaches based on passive methods cover a wide range of technologies and techniques from manual gamma-ray (γ-ray) intensity surveying with hand-held instruments to systems that make high-resolution three-dimensional imaging using the γ-ray emitted from the object being investigated. After specific sources of gamma-ray radiation are located, the radioactive isotopes in the sources can be identified using γ-ray spectroscopy. Active interrogation techniques with neutral particle beams provide possibilities of spatial imaging with isotope analysis. Neutron and gamma-ray beams are the current front runners for probes in active interrogation. Advancing the technologies for remote material assaying requires investments in basic science and engineering research and in education of scientists and technologists.

The main goals of the research programs at the Triangle Universities Nuclear Laboratory (TUNL) in the areas of nuclear security and nonproliferation are to contribute to the databases used in developing γ-ray beam based material analysis technologies and to educate physicists in scientific and technical areas relevant to this field. In this paper the photonuclear research program on actinide nuclei and the beam capabilities of the High Intensity Gamma-ray Source (HIγS) at TUNL are described.

2. Photonuclear Reaction Measurements on Actinides

Two distinct programs studying photonuclear reactions on actinides are underway at HIγS: nuclear resonance fluorescence (NRF) and polarized photofission with detection of the prompt emitted neutrons. In addition to adding to the nuclear database for technology development, the measurements made in these programs provide structure information on complex many-body systems and illustrate the dynamical response of heavy nuclei to electromagnetic waves.

2.1. NRF Measurements on Actinides

Gamma-ray beam interrogation with detection of γ-rays is based on NRF (γ, γ') measurements. In the NRF process the incident γ-ray excites low-spin states ($\Delta l \leq 2$) in the target nucleus. If the excitation energy is below the particle separation energy, then the excited nucleus decays by emission of γ-rays at energies that are characteristic to that nucleus. The standard configuration uses γ-ray detectors withbetter than 1% energy resolution. However, there are some measurement strategies that relax the energy resolution requirement of the γ-ray detection system and consequently allows for higher detection efficiency per cost (see article by C.T. Angell in these proceedings). For sufficient penetration through materials the incident γ-rays should have energy greater that about 2 MeV. The challenge is to find and characterize low-spin states in key nuclei at excitation energies above 2 MeV.

The NRF reaction is an effective mechanism for locating and characterizing nuclear states that can be excited by dipole and quadrupole γ-ray transitions from the ground state. An efficient approach to applying NRF techniques in nuclear structure studies is to perform a survey for dipole transitions over a broad energy range using a bremsstrahlung produced γ-ray beam followed by measurements at specific energies with a linearly polarized γ-ray beam with a narrow energy width. The latter measurements have higher sensitivity than those made with bremsstrahlung beams because of the reduction in the background and provide determinations of the type of transition, electric or magnetic, using the linear polarization of the beam.

During the last few years NRF measurements have been performed on several heavy nuclei over the energy range between 2 and 4 MeV at HIγS. In these experiments the integrated cross sections are measured and the branching ratios between transitions to the ground state and the first excited state are determined. In addition, for nuclei with a $J = 0^+$ ground state, the spin and parity of the excited states are determined. An example of recent work at HIγ S is the (γ, γ') measurements on ^{240}Pu which are described next.

Quiter *et al.* conducted a broad energy survey on ^{240}Pu using a bremsstrahlung beam in which they reported observing nine γ-ray transitions in the energy range from 2.1 to 2.8 MeV[1]. We followed up on their work to determine the spin and parities of the excited states associated with observed transitions using the linearly polarized and nearly mono-energetic γ-ray beam at HIγ S. Our measurements were performed between 1.95 and 2.95 MeV in 0.10-MeV steps. The γ-ray beam had an energy spread of about 5% fwhm and a time structure with 300 ps fwhm wide pulses that are separated by 180 ns. The time structure of the beam enables rejection of backgrounds due to the radioactivity of the sample, natural radiation in the room and cosmic radiation. The experimental setup used in our measurements is shown in Figure 1. The electric field vector of the beam is in the horizontal plane.

Examples of a time-of-flight and energy spectrum of γ-rays detected at θ= 90° in the horizontal detectors for the NRF measurements on ^{240}Pu are shown in Figure 2. The spectra were collected for an incident γ-ray beam energy of 2.55 MeV. The ^{240}Pu target was a cylindrical container of dimensions 19 mm dia. × 25 mm length with 4.6-g of PuO_2 powder inside. The spectra shown in Figure 2 were accumulated with about 5.5 hours of beam on target. The backgrounds in the energy spectrum were determined using time-of-flight techniques as shown in Figure 2. To determine the cross section using the non-uniform powder target required measurement of the density profile of the target. This measurement was perform using the γ-ray beam and beam profile imager at HIγS.

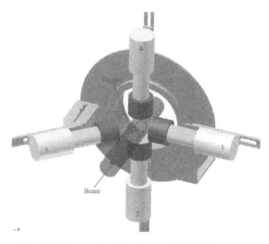

Fig. 1. 3D rendering of the standard experimental setup used for NRF measurements at HIγS. In this setup four 60% HPGe detectors are placed around the target at a scattering angle of θ = 90° with two of the detectors positioned with their center in the horizontal plane (referred to as horizontal detectors) and the other two detectors placed with their centers in a plane perpendicular to the horizontal plane (referred to as vertical detectors). In addition, to distinguish between M1 and E2 transitions, two more HPGe detectors (normally 25% efficient) are positioned at θ = 135° with their centers in the horizontal plane.

The NRF process is indicated in the schematic energy level diagram of ^{240}Pu in Figure 3. The nucleus is excited from the ground state ($J^\pi = 0^+$) to either a 1^+ state by an electric dipole (E1) transition or a 1^- state by a magnetic dipole (M1) transition. The γ-ray absorption is represented in the diagram by the upward pointing red arrow. The excited nucleus decays by the emission of γ-rays with an azimuthal angular dependence that is characteristic of the type of electromagnetic transition (E1 or M1). Measurement of the asymmetry in the number of γ-rays detected in the horizontal detectors versus those detected in the vertical detectors provides a direct determination of the transition type and gives the parity of the excited state. For example, for M1 transitions, all γ-rays will be emitted in the direction of the horizontal detectors, as is the case for the states with excitation energies of 2566 and 2578 keV shown in Figure 3. Our measurements confirmed the nine excited states observed by Quiter et al.[1] In addition, we discovered two additional J=1 states, one at 2444 keV and the other at 2834 keV. All observed states were determined to have spin and parity of $J^\pi = 1^-$. A characteristic feature of this nucleus and other actinides is the substantial fraction of decays to the first excited state ($J^\pi = 2^+$) relative to decays to the ground state. This attribute could potentially be used as a signature of actinides in active material analysis techniques. The transitions from the excited state to the ground state or the first excited state in ^{240}Pu are indicted in Figure 3 by the double arrays separated by 42.8 keV (i.e., the energy of the first excited state in ^{240}Pu).

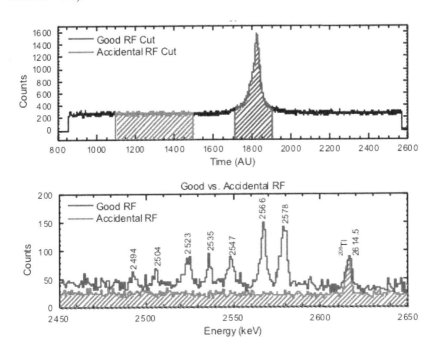

Fig. 2. Time-of-flight (top) and energy (bottom) spectra for γ-rays detected at θ = 90° in the horizontal plane. These spectra were accumulated for an incident γ-ray beam energy of 2.55 MeV. The shaded region around channel 1800 in the time-of-flight (tof) spectrum (blue) represents the detected γ-rays that are coordinated with incident γ-ray beam pulse on the target, and the shaded region from about channel 1050 to 1500 (red) identifies the detected γ-rays that are not time correlated with the incident beam pulse, i.e., the "accidental" counts. The counts in the latter region of the tof spectrum are due to background processes. The counts in the energy spectrum (bottom panel) that come from interactions of the γ-ray beam with the target are represented by the open spectrum (blue), and the shaded spectrum are background counts that are determined by the "accidental" region selected in the tof spectrum.

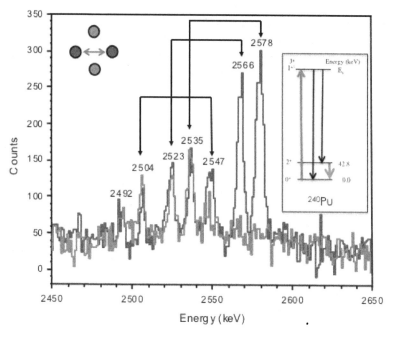

Fig. 3. The energy spectra of γ-rays detected at θ = 90° in the horizontal (blue) and vertical (red) detectors for the NRF measurements on ^{240}Pu. The centroid energy of the γ-ray beam used in the measurements was 2.55 MeV. The insert is a schematic of the energy level diagram indicating the NRF technique applied to ^{240}Pu.

2.2. NRF Measurements on Actinides

The other program that provides nuclear reaction data relevant to developing non-proliferation technologies is the study of polarized photofission. The goal of this program is to explore the possibility of using the asymmetry in the emitted prompt neutron yield from polarized photofission as a technique for discerning between fissile and nonfissile material as the basis of a technology for remote assaying. In this study nuclear fission is induced with linearly polarized γ-ray beams with about a 5% energy spread, and the yields of the prompt neutrons emitted from the fission of the target nucleus in the plane of the beam polarization and those emitted perpendicular to the polarization plane are measured. A schematic diagram of the concept is shown in Figure 4. Measurements on ^{235}U, ^{238}U, ^{239}P and ^{232}Th have been recently reported by Mueller et al.[2] The measurements were performed over the incident γ-ray beam energy range of about 5.7 to 6.2 MeV to cover the energy region from just below the fission threshold to just below the (γ, n) threshold. The results reported by Mueller *et al.* are summarized in the plot of the neutron asymmetry P_n as function of the incident γ-ray beam energy (E_γ) shown in Figure 5.

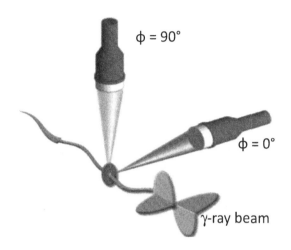

Fig. 4. Schematic of the polarized photofission experiment setup.

The neutron yields are integrated for detected neutrons with energies greater than 1.5 MeV. The curves shown in Figure 5 are simulations based on a model of the fission process. These results suggest a pattern that R_n has a value of about one for fissile nuclei whereas non-fissile actinides have a ratio greater than one.

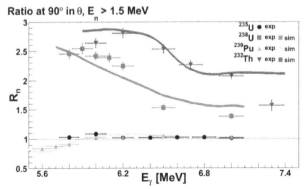

Fig. 5. The measured P_n at θ = 90° for E_n > 1.5 MeV as a function of E_γ (from the paper by Mueller et al.[2]).

3. HIγ S

The γ–ray beams at HIγS are produced by Compton backscattering of photons from electron bunches circulating inside the optical cavity of a storage-ring based Free Electron Laser (FEL). The high intensity of this source, about 1000 γ/s/keV, is mostly due to the combination of the high intra-cavity optical power and the high average beam current in the storage ring (about 100 mA). The layout of the facility is shown in Figure 6. The FEL consists of electromagnetic undulators that are installed in one of the straight sections of the 1.2-GeV racetrack-shaped storage ring. The undulators form the active elements of optical klystron (OK) FELs with a long optical resonator of 53.7 m in length (mirror-to-mirror distance). The electron accelerator drivers consist of a 180-MeV linac pre-injector, a booster injector of 180 MeV to 1.15 GeV energy range and a race-track shaped storage ring that has an energy range of 250 MeV to 1.15 GeV. The storage ring circumference is about 108 m.

The γ-ray beams at HIγS are nearly mono-energetic and highly polarized (linear or circular). The beam energy is tunable by adjusting the electron energy and magnetic field strength in the undulators. The γ-ray beam energy range is from 1 to 100 MeV. The energy spread of the γ-ray beam is selectable to about 1% (FWHM) by collimation. The γ-ray beam specifications and information about the HIγS facility are available on the HIγS web site: http://www.tunl.duke.edu/higs/. The technologies employed to produce γ-rays at HIγS are robust and enables delivery of beam to experiments with greater than 95% reliability.

An upgrade to increase the γ-ray beam intensity at HIγS by about two orders of magnitude at energies below about 10 MeV is under consideration. The upgrade concept is referred to as HIγS2. The HIγ ∴ S2 will be a complementary γ-ray source to the existing FEL driven source. This next-generation Compton γ-ray source will add to the γ-ray capabilities at the HIγS in the energy range of about 2 to 12 MeV. Driven by a high average-power Fabry-Perot resonator, the HIγS2 will produce γ-ray beams with intensities more than two orders of magnitude and with better energy resolution than the current source at HIγS. The HIγS2 will offer highly polarized beams of circular polarization with the capability of rapid helicity switching. The configuration strategy is to maintain the current capabilities of the existing FEL driven source at HIγS with additional capabilities provided by HIγS2 at low energies.

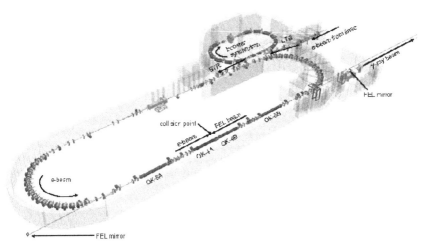

Fig. 6. Layout of the HIγS accelerator systems.

Acknowledgments

This work is supported in part by the U.S. Department of Homeland Security grant number US-DHS-2012-DN-077-ARI062-02 and the U.S. Department of Energy National Nuclear Security Administration grant number DE-NA0001838. I thank my colleagues at Duke University, Professors Henry Weller and Ying Wu, for the materials they provided for this paper and for their discussions on the topics presented.

References

1. B.J. Quiter *et al.*, *Phys. Rev. C* **86**, 034307 (2012).
2. J.M Mueller *et al.*, *Phys. Rev. C* **85**, 014605 (2012).

Dipole Strength Around the Particle Threshold

D. Savran

ExtreMe Matter Institute EMMI and Research Division,
GSI Helmholtzzentrum für Schwerionenforschung GmbH, Darmstadt, Germany
and
Frankfurt Institute for Advanced Studies FIAS, Frankfurt am Main, Germany
d.savran@gsi.de

In the low-energy part of the electric dipole strength of atomic nuclei an addition structure has been identified beside the well-known Giant Dipole Resonance, which is usually denoted as Pygmy Dipole Resonance (PDR). The PDR has attracted strong interest in the last years in nuclear physics. Different experimental approaches have been used in the last decade in order to investigate this new interesting nuclear excitation mode. Among the different experiments the nuclear resonance fluorescence (NRF) method has played a pioneering role in the research on the PDR.

Keywords: Pygmy dipole resonance; NRF; low-lying E1 strength.

1. Introduction

In atomic nuclei nearly the complete electric dipole (E1) strength is concentrated in the well-known Isovector Electric Dipole Giant Resonance (IVGDR). This collective excitation mode can be understood as an out-of-phase oscillation of the protons and neutrons (in a macroscopic interpretation) and has been studied in the past in various experiments, see e.g.[1] and references therein.

In the last years a structure of E1 strength located at lower excitation energies has attracted considerable interest in nuclear physics. In many nuclei additional strength is experimentally found well below the IVGDR, which is mostly denoted as Pygmy Dipole Resonance (PDR) or Pygmy Dipole Strength. The strength located in the PDR is usually on the order of a (few) percent of the IVGDR, while the centroid energy is in the vicinity of the particle emission thresholds.

Even though the PDR exhausts only a rather small fraction of the corresponding energy weighted sum rule (EWSR), it is of particular interest to test details of modern microscopic models which describe the E1 response of atomic nuclei[2]. In addition it has gained interest in the last decade due to its possible connections to the neutron skin of atomic nuclei and properties of nuclear matter[3-7] as well as reaction rates in the synthesis of heavy elements[8-10]. However, the microscopic structure of the PDR, its collectivity and the robustness of the connections to other properties mentioned above are a matter of ongoing discussions.

Experimentally the situation is not conclusive regarding the systematics of the PDR and data providing insight into its structure is still rather scarce. We have recently presented the available experimental data on the PDR in our review[11] describing in detail also the experimental approaches and their limitations. Special attention is made in this review also on the NRF method using different photon beams and its application to investigate low-lying E1 strength below the particle threshold. For a detailed discussion I thus refere to[11].

2. The strength of the PDR

Information on the (total) strength of the PDR can be obtained best in real- or virtual-photon induced reactions. On stable nuclei the nuclear resonance fluorescence (NRF) method has been most widely used to gain information on the E1 strength distribution below the neutron separation energy. For these experiments photon beams produced via bremsstrahlung[12,13] and laser Compton backscattering (LCB)[14] have been used. Different approaches in the analysis have been used taking into account either only the information of isolated peaks resulting in model-independent results but accepting a limited sensitivity (see e.g.[15]) or including the full spectra to extract the full strength but accepting the need of statistical model assumptions in the analysis and thus model dependent results (see e.g.[16]). For a detailed discussion of the different approaches used in NRF experiments including their individual advantages and drawbacks I refere to the review[11] and the references given above.

In the energy regionabove the neutron threshold LCB photons have been used to measure the (γ, n) cross section as a function of the excitation energy (see e.g.[17,18]). Just recently a new approach has been proven to result in the extraction of the full E1 response independent of particle thresholds by the means of Coulomb excitation in the (p,p') reaction[19,20], which however requires a separation from the M1 response via a multipole decomposition analysis.

For unstable nuclei the available approaches are much more limited. So

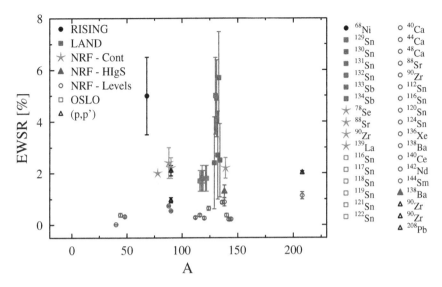

Fig. 1. Reported exhaustion of the PDR in various nuclei extracted from experiments using different experimental approaches. Figure taken from [11].

far the E1 response in medium-heavy to heavy nuclei has been studied via Coulomb excitation in inverse kinematics in the region of ^{130,132}Sn[4,21] and in ^{68}Ni[22]. Both experiments were sensitive to excitation energies above the corresponding particle thresholds. Thus, deriving a consistent systematic picture on the total strength of the PDR together with the results from stable nuclei is very difficult.

Figure 1 summarizes the reported EWSR values for a variety of nuclei of different masses (see [11] for details). However, one needs to keep in mind that the values are not derived in a consistent way, i.e. in some cases contributions of the low energy GDR tail were subtracted and different energy regions have been chosen for the integration region. This summary therefore can only provide a first idea on the strength of the PDR.

3. The decay properties of the PDR

A property which can give further insight into the structure of the E1 strength, especially the coupling of complex configurations and the damping mechanism, is the decay pattern of the PDR. So far this property was not possible to directly access, since the branching transitions to individual low-lying excited states are small and a very sensitive experimental method is needed. Often the concept of γ-ray strength functions within the statistical model is used to describe the decay of excited states in the PDR energy region. Partly this concept is also used in the analysis of experimental data do derive information on the total strength of the PDR. An experimental determination of branching ratios and in general the decay pattern of the low-lying E1 strength is therefore of principle interest.

Standard NRF experiments are not sensitive enough to investigate the weak branching transitions to low-lying excited states, even when using a mono-energetic LCB photon beam. At the same time these NRF experiments have shown indirectly that the integrated intensity going into these so-called inelastic transitions (i.e. any transition not going directly back to the ground state) cannot be neglected[23,24]. The method of γ-γ coincidence experiments has been shown to be very sensitive even to very weak transitions. However, so far this method has only been used in particle-induced reactions, which do not sufficiently excite the $J^{\pi} = 1^{-}$ states of interest. Therefore, the combination of γ-γ coincidence spectroscopy and the NRF reaction using a mono-energetic photon beam from LCB offers the best possible experimental approach to access the branching pattern of the PDR.

To realize this approach the γ^3 collaboration has installed a new high-efficiency γ-spectroscopy setup at the High Intensity γ-ray Source (HIγS) at Duke University, the new γ^3 setup[25]. This setup combines high-efficiency large LaBr Detectors with high-resolution HPGe detectors. The concept is illustrated in Fig. 2. The high efficiency of the LaBr array allows to gate on the decay of the low-lying state (green) and investigate the feeding transitions (red) with high resolution using the HPGe array. Since the photon beam is mono-energetic all transition energies are uniquely defined in a two-step decay cascade, which strongly reduces the background and, thus, provides the necessary sensitivity. In a first commissioning experiment this has been demonstrated for the case of ^{32}S, where an improvement of the peak-to-background ratio of about 50 was achieved[25].

The first experiments to investigate the PDR in the Z=50 and N=82 mass regions have been performed in the end of 2012. The first results show that the sensitivity is high enough in order to investigate directly the branching ratios into the first excited states. This new data will be confronted to microscopic calculations as well as the statistical model.

In summary, the experiments performed so far to investigate the PDR have shown, that this new excitation mode is a general phenomenon of medium-heavy to heavy nuclei. In many aspects the available experimental

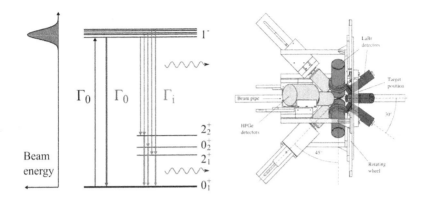

Fig. 2.   Concept of the γ-γ coincidence technique in combination with the monoenergetic photon beam at HIγS. The right part shows a schematic drawing of the new γ^3 setup consisting of HPGe (green) and LaBr detectors (red and blue)[25].

data is still incomplete and the open questions need to be answered by the experiments which are currently under way.

This work is supported by the Alliance Program of the Helmholtz Association (HA216/EMMI).

References

1. M. N. Harakeh and A. van der Woude, *Giant Resonances*, Oxford University Press (2001).
2. N. Paar, D. Vretenar, E. Khan, and G. Colò, Rep. Prog. Phys. 70 (2007) 691.
3. J. Piekarewicz, Phys. Rev. C 73 (2006) 044325.
4. A. Klimkiewicz, N. Paar, P. Adrich, *et al.*, Phys. Rev. C 76 (2007) 051603.
5. N. Tsoneva and H. Lenske, Phys. Rev. C 77 (2008) 024321.
6. J. Piekarewicz, Phys. Rev. C 83 (2011) 034319.
7. D. Vretenar, Y. F. Niu, N. Paar, and J. Meng, Phys. Rev. C 85 (2012) 044317.
8. S. Goriely, Phys. Lett. B 436 (1998) 10.
9. S. Goriely, E. Khan, and M. Samyn, Nucl. Phys. A739 (2004) 331.

10. E. Litvinova, H.P. Loens, K. Langanke, G. Martinez-Pinedo, T. Rauscher, P. Ring, F.-K. Thielemann, and V. Tselyaev, Nucl. Phys. 823 (2009) 26.

11. D. Savran, T. Aumann, and A. Zilges, Prog. Part. Nucl. Phys. 70 (2013) 210.

12. R. Schwengner, R. Beyer, F. Dönau, et al., Nucl. Instr. and Meth. A 555 (2005)(1-2) 211 .

13. K. Sonnabend, D. Savran, J. Beller, et al., Nucl. Instr. and Meth. Phys. Res. A 640 (2011) 6.

14. H. R. Weller, M. W. Ahmed, H. Gao, W. Tornow, Y. K. Wu, M. Gai, and R. Miskimen, Prog. Part. Nucl. Phys. 62 (2009)(1) 257 .

15. D. Savran, M. Elvers, J. Endres, et al., Phys. Rev. C 84 (2011) 024326.

16. R. Schwengner, G. Rusev, N. Tsoneva, et al., Phys. Rev. C 78 (2008) 064314.

17. H. Utsunomiya, S. Goriely, H. Akimune, et al., Phys. Rev. C 81 (2010) 035801.

18. H. Utsunomiya, S. Goriely, M. Kamata, et al., Phys. Rev. C 84 (2011) 055805.

19. A. Tamii, I. Poltoratska, P. von Neumann-Cosel, et al., Phys. Rev. Lett. 107 (2011) 062502.

20. C. Iwamoto, H. Utsunomiya, A. Tamii, et al., Phys. Rev. Lett. 108 (2012) 262501.

21. P. Adrich, A. Klimkiewicz, M. Fallot, et al., Phys. Rev. Lett. 95 (2005) 132501.

22. O. Wieland, A. Bracco, F. Camera, et al., Phys. Rev. Lett. 102 (2009) 092502.

23. A. P. Tonchev, S. L. Hammond, J. H. Kelley, E. Kwan, H. Lenske, G. Rusev, W. Tornow, and N. Tsoneva, Phys. Rev. Lett. 104 (2010) 072501.

24. J. Isaak, D. Savran, M. Krticka, et al., Phys. Lett. B 727 (2013) 361.

25. B. Löher, V. Derya, T. Aumann, et al., Nucl. Instr. and Meth. Phys. Res. A 723 (2013) 136.

Resonant Photonuclear Reactions for Neutrino Nuclear Responses and Nuclear Isotope Detections

H. Ejiri

RCNP, Osaka University, Ibaraki, Osaka, 567-0047,
Japan, Nuclear Science, CTU, Prague, Czech republic
ejirir@rcnp.osaka-u.ac.jp
www.rcnp.osaka-u.ac.jp.~ejiri

This is a brief report onrecent studies of photonuclear reactions by using medium energy photons produced by laser photons scattered off GeV electrons. The medium energy photonuclear excitations of IAR (isobaric analogue resonance) are used to study neutrino nuclear responses relevant to double beta decays and astro-neutrinos. Gamma rays following medium-energy resonant photonuclear reactions are used for high-sensitivity non-destructive nuclear isotope detections.

Keywords: Photonuclear reaction; laser photons scattered off GeV electrons; neutrino nuclear responses; non-destructive nuclear isotope detection.

1. Resonant photonuclear reactions by medium energy photons

Medium energy photons in a range of E=12-30 MeVare obtained from laser photons back-scattered off GeV electrons. They are very useful for resonant photonuclear reactions. We report briefly recent works on resonant photonuclear reactions by the medium energy photons for studies of neutrino nuclear responses associated with double beta decays and astro neutrinos[1] and for non-destructive detections of nuclear isotopes of astro-nuclear, geological and historical interests[2]. Details are given in the articles[1,2] and refs. therein.

2. Resonant photonuclear reactions for neutrino nuclear responses

Medium-energy polarized photons obtained from laser photons scattered off GeV electrons provide a unique opportunity to study neutrino/weak nuclear responses involved in astro-neutrino reactions and double beta decays[3-5].

The neutrino nuclear responses are studied by using beta decays and muon captures via weak interactions, photonuclear and electron reactions via electromagnetic (EM) interactions and nuclear reactions via nuclear interactions, as shown in Fig. 1.

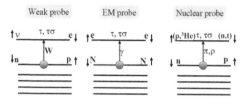

Fig. 1. Neutrino nuclear response studies by using lepton probes with weak interactions, photon and electron probes with electromagnetic interactions and nuclear probes with nuclear interaction[3-5].

Electromagnetic interactions have similar spin isospin operators as weak interactions. Then electric and magnetic photonuclear reactions are used to study neutrino nuclear responses for vector and axial vector weak interactions. Charged current weak responses for ground and excited states are studied by using photonuclear reactions through isobaric analog resonances (IAR) of those states as shown in Fig.2, while neutral current weak responses for excited states are studied by usingphoto nuclear reactions through the excited states.

The weak interaction strengths are obtained from the photonuclear reaction cross-sections, and the spin and parity of the state are derived from the angular correlations of the photo nuclear reactions as shown in Figs. 3 and 4[1]. The neutrino responses are crucial for the neutrino studies in nuclei.

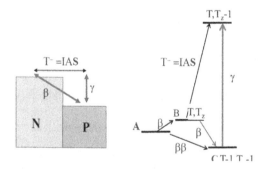

Fig.2. Reaction and transition schemes of single beta and double beta decays and photonuclear reaction through IAR[1].

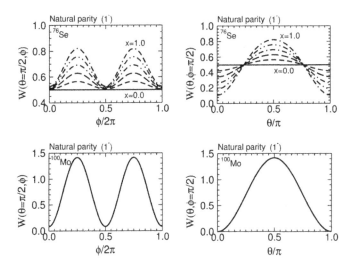

Fig. 3. Top. Azimuthal and polar angular distributions of neutrons from photonuclear reactions on ^{82}Se via a natural parity 1⁻ IAR with x being the fraction of d-wave neutron. Bottom: angular distributions for ^{100}Mo[1].

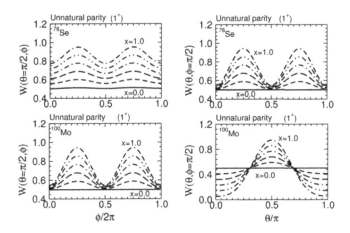

Fig. 4. Top. Azimuthal and polar angular distributions of neutrons from photonuclear reactions on ^{82}Se via an unnatural parity 1⁺ IAR with x being the fraction of f-wave neutron. Bottom: angular distributions for ^{100}Mo with x being the fraction of d-wave neutron[1].

3. Resonant photonuclear isotope detection

Resonant photonuclear isotope detection (RPID) is a new non-destructive high-sensitive detection of nuclear isotopes[2]. Medium energy 12-16 MeV photons from laser photons back scattered off GeV electrons are used for photonuclear (γ,n) reactions via the E1 giant resonance, and γ rays characteristic of the reaction products are measured by means of high-sensitive Ge detectors. RPID is shown to be used to detect stable and radioactive isotopes of the orders of μ gr – n gr and ppm—ppb impurities for studying nuclear isotopes of physics, geological and historical interests, and also to detect fission products and U/Th isotopes as well.

The cross section of the photonuclear (γ, n) reaction via the E1 giant resonance is around 20-30 % of the geometrical cross section for all nuclear isotopes. Intense photons in a wide band $\Delta E{\sim}4$ MeV are used to excite preferentially the E1 giant resonance to produce radioactive nuclei. Then delayed γ rays characteristic of the reaction products are efficiently measured by using low background (BG) high energy-resolution Ge detectors with NaI active-shields as shown in Fig.5. Since the laser-electron photons are well collimated, one can study impurities in a small sample, and their position can be identified by scanning the sample.

The detection sensitivity is evaluated for three cases of A: low BG singles γ measurement, B: low BG coincidence measurement of cascade γ rays to reduce BGs, and C: singles measurement in case of BG γ rays associated with reaction products from other isotopes in the sample. They are shown in Fig. 6.

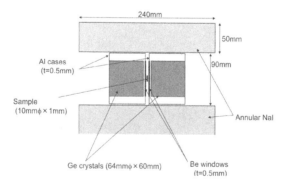

Fig. 5. Low BG γ detection system. It consists of two high-efficiency Ge detectors with annular NaI active shields[2].

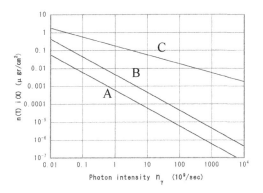

Fig. 6. RPID sensitivities. A: Low BG singles measurement. B : Low BG coincidence measurements. C: Singles measurements with BGs from photonuclear reaction products of other isotopes[2].

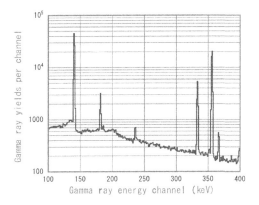

Fig. 7. Gamma ray spectrum from photonuclear reactions on Mo sample with 7.6% Au. The 140.5 and 181 keV γ rays are from ^{99}Mo decays. The 333 and 356 keV γ rays are from ^{196}Au decays[2].

Medium energy 12-16 MeV photons from laser photons back scattered off the NewSUBARU 1 GeV electrons were used to demonstrate the feasibility of RPID. The photon intensity was about 0.9×10^6/sec. The energy spectrum is shown in Fig.7. Gamma rays from natural Mo and Au isotopes are clearly seen. The observed spectrum leads to the sensitivity of around 4 ppb fordetecting Au impurity in a Mo sample in case of the high intensity photon source of 10^9/sec.

Resonant photonuclear reactions are used to produce ^{99}Mo/^{99}Tc and other medial RIs and to transmute long-lived fission products. High intensity $10^{12\text{-}15}$ photons /sec produce specific nuclear isotopes with the rate of $10^{10\text{-}13}$/sec and the density of $0.05 - 50$ G Bq/mg[6].

Recently we have shown that γ rays from RI reaction products of muon capture reactions are used to study neutrino nuclear responses and for nuclear isotope detection[7]. MCID (muon capture isotope detection)[7] is complementary to the present RPID[2].

References

1. H. Ejiri, A. Titov, M. Bosewel and A. Young, *Phys. Rev.* C **88**, 054610 (2013).
2. H. Ejiri *Phys. Rev. ST,* **15**, 024701 (2012).
3. H. Ejiri, *Prog. Nucl. Particle Phys.* **64**, 249 (2010).
4. J. Vergados, H. Ejiri and F. Simkovic, *Rep. Progress Phys.* **75**, 106301 (2012).
5. H. Ejiri, *Phys. Rep.* C **338**, 265 (2000).
6. H. Ejiri, *J. Phys. Soc. Jpn.* **80**, 094202 (2011).
7. H. Ejiri, I. H. Hashim, Y. Hino, Y. Kuno, Y. Matsumoto, K. Minomiya, H. Sakamoto, A. Sato, T. Shima, A. Shinohara, K. Takahisa, and N.H. Tran, *J. Phys. Soc. Jpn.* **82**, 044202 (2013).

Non-Destructive Testing for Control of Radioactive Waste Package

S. Plumeri

Andra (French national radioactive waste management agency)
1-7 rue Jean Monnet, 92298 Chatenay-Malabry, France
stephane.plumeri@andra.fr
www.andra.fr

F. Carrel

CEA, LIST, Sensors and Electronic Architectures, Gif-sur-Yvette, F-91191, France
frederick.carrel@cea.fr

Characterization and control of radioactive waste packagesare important issues in the management of a radioactive waste repository. Therefore, Andra performs quality control inspection on radwaste package before disposal to ensure the compliance of the radwaste characteristics with Andra waste disposal specifications and to check the consistency between Andra measurements results and producer declared properties. Objectives of this quality control are: assessment and improvement of producer radwaste packages quality mastery, guarantee of the radwaste disposal safety, maintain of the public confidence. To control radiological characteristics of radwaste package, non-destructive passive methods (gamma spectrometry and neutrons counting) are commonly used. These passive methods may not be sufficient, for instance to control the mass of fissile material contained inside radwaste package. This is particularly true for large concrete hull of heterogeneous radwaste containing several actinides mixed with fission products like ^{137}Cs. Non-destructive active methods, like measurement of photofission delayed neutrons, allow to quantify the global mass of actinides and is a promising method to quantify mass of fissile material. Andra has performed different non-destructive measurements on concrete intermediate-level short lived nuclear waste (ILW-SL) package to control its nuclear material content. These tests have allowed Andra to have a first evaluation of the performance of photofission delayed neutron measurement and to identify development needed to have a reliable method, especially for fissile material mass control in intermediate-level long lived waste package.

Keywords: Waste management; quality control; photofission; actinides mass.

1. Context

The French deep geological disposal facility for radioactive wastes, named Cigéo (in project) would serve as a repository for highly radioactive long-lived wastes generated by France's current fleet of nuclear power plants, as well as from reprocessing of spent fuel from these same plants. Although high-level (HLW) and intermediate-level long-lived (ILW-LL) wastes represent a small volume of the wastes generated (3% of the volume of existing radioactive wastes), they account for nearly all the radioactivity of radioactive wastes (around 99%). A 500 meters below ground in an impermeable argillaceous rock formation (Callovo-Oxfordian argillite) able to contain radioactivity over a very long period (see Figure 1) is studied to get the waste.

Figure 1: The Cigéo project

2. Radioactive Waste Quality Control

Characterization and control of radioactive waste packages are important in the management of a radioactive waste repository. Therefore, Andra performs quality control inspection on radwaste package before disposal to ensure the compliance of the radwaste characteristics with Andra waste disposal specifications and to check the consistency between Andra measurements results and producer declared properties. Objectives of this quality control are: assessment and improvement of producer radwaste packages quality mastery, guarantee of the radwaste disposal safety, maintain of the public confidence.

3. Requirements

To control radiological characteristics of radwaste package, non-destructive passive methods (gamma spectrometry and neutrons counting) are commonly used. These passive methods may not be sufficient, for instance to control the mass of fissile material contained inside radwaste package. This is particularly true for large concrete hull (1-2 m^3) of heterogeneous radwaste containing several actinides mixed with fission products like ^{137}Cs. The mass of this kind of package is a few tons and it contains a total βγ activity of few TBq (mean value ≈ 3 TBq) and a maximal Pu mass of 70 g. The challenge is to perform the quantification of the fissile material mass inside this radioactive waste package (see Fig. 2), while controlling measurement uncertainties.

Figure 2: Example of ILW-LL package

4. Studies on Photofission Based Methods

Non-destructive active methods, like measurement of photofission delayed neutrons, allow to quantify the global mass of actinides and is a promising method to quantify mass of fissile material [1]. This technique employs intense Bremsstrahlung photon beam with energy above 6 MeV to induce photofission of actinides.

MCNPX simulations have shown adequacy between detection limits of the method and mean actinide mass in ILW-LL packages. It has also shown that uncertainty on hydrogen mass inside the package implies a large uncertainty on actinide mass evaluation.

Measurements of photofission delayed neutrons and gammas have been performed on a real ILW-SL package using the SAPHIR facility (CEA LIST, Saclay site). The radioactive waste package used for these tests has the same configuration and geometry as ILW-LL packagebut witha quantity of fissile material lower than what expected in ILW-LL (685 mg of ^{239}Pu). Quantification of Pu by passive neutrons counting and delayed induced neutrons counting wereconsistent but large uncertainties were evaluated mainly due to uncertainties on background noise sources and matrix effects.

5. Conclusion

For some kind of radioactive waste package (large concrete volume), measurement of gammas and neutrons induced by photofission is a promising method to quantify actinides mass and this has been validated by simulation and first tests on ILW-SL package. Systematic estimation on background noise has to be performed to obtain a reliable measurement of actinide mass. To completely fulfill requirements of quality control of ILW-LL package, performance of photofission measurement has to be measured in term of discrimination of non-fissile U from other fissile actinides. Delayed gamma-ray spectrometry is a method that may be used to achieve this kind of measurement [2].

Acknowledgments

This work was supported in the frame of R&D studies conducted jointly by CEA and Andra for the characterization of intermediate-level long-lived radioactive waste.

References

1. F. Carrel, M. Agelou, M. Gmar, F. Lainé, B. Poumarède, B. Rattoni, *Measurement of plutonium in large concrete radioactive waste packages by photon activation analysis*, IEEE Transactions on Nuclear Science, vol. 57, pp. 3687–3693, 2010.
2. F. Carrel, M. Agelou, M. Gmar, F. Lainé, J. Loridon, J.L. Ma, C. Passard, B. Poumarède, *Identification and Differentiation of Actinides Inside Nuclear Waste Packages by Measurement of Delayed Gammas*, IEEE Transactions on Nuclear Science, vol. 57, pp. 2862–2871, 2010.

Development of First Responders Equipment at RN Incident Sites[*]

K. Tsuchiya, K. Kuroki, K. Kurosawa and N. Akiba

Physics Section, National Research Institute of Police Science, 6-3-1 Kashiwanoha, Kashiwa, 277-0882, Japan

On site categorization and collection of radioactive and nuclear materials are required at radiological and nuclear incident site. We are developing portable equipment and radiation protection for radiological emergency response team to carry out emergency missions safely at the incident sites. In this report, we review radiation monitoring system including wireless dosimeter system and neutron shield with water developed in our institute. Also the development of fast-neutron directional detector with a micro pattern gas detector is described.

1. Introduction

Since the beginning of the 2000s, a radiological or nuclear (RN) threat is significantly increased and homeland security (national security) systems to support the domestic nuclear detection have been developed in the world. Nuclear forensics is the technical means by which nuclear materials or radioactive materials are characterized as to composition, physical condition, age, provenance, and history [1, 2]. The categorization of nuclear materials to a significant degree by measurements from portable instruments on-site are required. Also collection method of the nuclear materials or post-explosion debris for RN incident such as a dirty bomb or a radiological dispersal device (RDD) is needed for obtaining samples for laboratory analysis.

Several scenarios for RN terrorism threat in Japan are conceivable. One of them is RDD, named as "dirty bomb". This combines a conventional explosive with radioactive material and disperse radioisotopes such as Cs-137, Co-60, or Ir-192 which are widely used in industrial and medical field. Special nuclear materials (SNM) defined as any plutonium isotopes and enriched uranium are also afraid as source of RDD. After explosion, contamination area is expected to be wide, and radiological debris will spread in the environments. Actual nuclear accident is helpful in considering scenarios on RN incidents. Nuclear criticality

---

[*] This work was supported by special research in National Police Agency andby JSPS KAKENHI Grant Number 21760717.

accident at Tokai, Japan was occurred on September, 1999. Two workers near uranium solution were dead. Criticality lasted for 20 hours, as a result, several rescue members, workers, and over 100 public persons were exposed to neutron radiation. If the terrorist use the same type devices, the high dose-rate area is expected to be widespread, because the fast neutron from nuclear materials can penetrate air over 200m.In Fukushima nuclear accident, first responders had worked to prevent the loss of cooling water. The gamma-ray dose-rate at the site was over 300mSv/h (30rem/h). We must deal with such a high dose field in the case of RN incident.

2. Neutron shield

In case of nuclear criticality devices, a fission chain reaction is developed and a nuclear excursion with a lethally strong emission of neutrons and gamma-rays are expected. Considering radiation damage, major harmful radiation is neutrons, especially fast neutrons. We have developed a neutron protection shield shown in Fig.1. We selected water as shield material for reducing fast neutrons. Light element is effective for neutron slowdown. Water can be easily obtained at the site and it is transparent. A long tong are attached at the center of the shield and sampling RN source or debris from behind the shield is available. The water shield is mounted on electric cart. A container made of lead is also attached to the cart. A thickness of prototype neutron shield is 10cm water and 2cm acrylic resin, by which neutron dose is reduced to 1/3 and gamma dose (Co-60) is reduced to 1/2.

Fig. 1. Neutron protection shield with water.

3. Dose control

Radiation monitoring vehicles were deployed in each prefecture, where nuclear plants were located, after the Tokai criticality accident. We integrated the functional devices of the radiation monitoring vehicles into trolley bag size. Fig. 2 showed a trolley bag type radiation measurement system. This system contains10 wireless network dosimeters for manage the individual dose at back-up force site, 3 types of hand-held detectors for dose rate measurement and radiological nuclide identification, and air sampler of aerosol for checking alpha emitter. Wireless network dosimeters developed by Fuji Electric were optimized for first responders. The wireless network range is about 100m at open site. They can reduce operation time by real time monitoring of neutrons and gamma-ray dose. We have evaluated active dosimeters that measure neutrons precisely under high dose-rate field using the Transient Experiments Criticality Facility (TRACY) at the Japan Atomic Energy Agency (JAEA). This is a reactor which can realize the nuclear criticality using 10% U-235 enriched uranyl nitrate solution as fuel [3]. Tested are real-time personal neutron dosimeters of different types, NRG13 (Fuji Electric, Japan)[4], ADM-353 (Hitachi-Aloka, Japan) [5] and DMC2000GN (MGP, United States)[6]. These solid state dosimeters have digital displays of dose, and a warning function using light, sound and/or vibration. The dosimeters were attached on a $30 \times 30 \times 15 \mathrm{cm}^3$ phantom. We irradiated them in two condition (Fig. 3), direct irradiation and irradiation with water shield (t=30cm). The dose indicated by ADM-353 was 2 times larger than that of NRG13 in director radiation condition, whereas the dose indicated by ADM-353 was over 8 times larger than that of NRG13 in water shield condition (Fig.4). They are caused by the energy responses of dosimeters and neutron spectra at the location. The details are described elsewhere [7]. We confirmed they worked over 6S v/h (600 rem/h) under neutron and gamma-ray irradiation using criticality field. We also confirmed the active dosimeters can work after prompt criticality.

Fig. 2.Trolley bag type radiation measurement system (left) and wireless network dosimeters (right).

Fig. 3. Neutron irradiation for active dosimeters using TRACY. Direct irradiation (left) and water shield (t=30cm) (right).

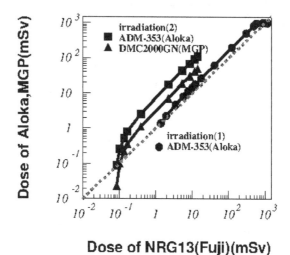

Fig. 4. Comparison between dose indicated by each neutron dosimeters, ADM-353, DMC2000GN, and NRG13. Irradiation (1) shows results of direct irradiation and irradiation (2) shows results of water-shielded condition (dosimeters were set behind a 30cm-thick water shield).

4. Nuclear material search

Stand-off detection of SNM is useful for first responders to a nuclear emergency. Gamma-ray imaging methods developed in medical, high-energy physics, and astronomy are applied for searching SNM. These methods are applied for uranium, which gamma-rays above 1MeV (1001 keV and 2614 keV) are good signatures for natural and depleted uranium. However, the gamma-ray signature from plutonium are 60 keV (Am-241) and 375/414 keV (Pu-239), which are relatively easy to shield with high-Z materials. We selected fast neutrons which penetrate high-Z shield as radiation for stand-off detection. Mean free path of fast neutrons in air is above 200m and the background rate from secondary products of cosmic-raysis low (about 1/50 of thermal neutrons on ground level). Their merit let us select a passive method for fission neutrons as one method of nuclear material search.

We have been developing a fast neutron directional detector with a micro-pixel gas chamber (μ-PIC). The μ-PIC is a micro-pattern gaseous imaging detector manufactured with printed circuit board technology. A schematic structure of the μ-PIC is shown in Fig. 1 of [8]. Anode and cathode strips are formed orthogonally on a polyimide substrate with a pitch of 400 μm. Anodepixels are located in each center of a cathode hole through the 100μm thick substrate using electric plating process. Time-resolved imaging for X-rays [9] and for thermal neutron [10] were achieved.

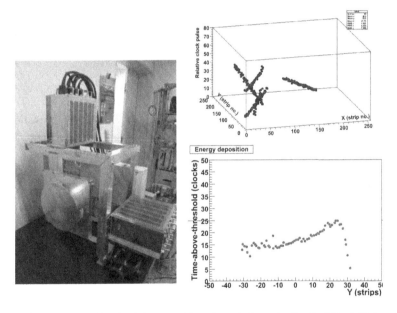

Fig. 5. Fast neutron directional detector with a micro-pixel gas chamber (μ-PIC). (left) detector and data acquisition system. (right) 3D-track of recoil protons and Bragg curve of a proton scattered by fast neutron.

Our method uses elastic scattering of neutrons with atomic nuclei in an Ar/CH₄gas mixture within a vessel. By measuring the 3D-track of recoil protons in a time projection chamber (TPC) based on μ-PIC, we can determine the direction of incident neutrons. The proof of concept for 3D-tracking of recoil proton was carried out using Cf-252 (Fig. 5). The direction was determined by gradient of energy deposit (Bragg curve) for recoil proton. This system have low background for neutrons and ability to distinguish neutron signal (recoil proton track) and gamma-ray background (electron track). High signal to noise ratio and large volume of detector are expected to cover low efficiency owe to small cross section of elastic scattering. Volume of prototype detector shown in Fig.4 is $10 \times 10 \times 5$ cm^3. If large TPC ($30 \times 30 \times 30$cm^3) is manufactured, we can detect weapon-grades Pu (critical mass 6kg) at a distance of 10m.

5. Summary

We have developed compact radiation monitoring system for first responders. The wireless network active dosimeters worked in high dose rate (Neutron and gamma-ray field over 6Sv/h) and after prompt criticality. The value of the neutron dose measured by each active dosimeters is 8 times different in the same irradiation condition, especially using water shield. Also we have been developing fast neutron directional device and verification of fast neutron detection using elastic scattering with TPC was done.

Acknowledgments

The authors would like to thank TRACY group in JAEA for providing irradiation field in our evaluation test of neutron dosimeters. We thank Dr. K. Tonoike for the support for this work. Also we wish to thank the members of Tanimori group in Kyoto University for development offast neutron directional detector and the members of neutron standard group in National Institute of Advanced Industrial Science and Tecnology (AIST) for the help of neutron dosimetry.

References

1. K. Mayer, M. Wallenius, and I. Ray, Analyst 130 433–441 (2005).
2. IAEA Nuclear Security Series 2 "Nuclear Forensics Support" (2006).
3. Y. Yamane, K. Nakajima, K. Ogawa, E. Aizawa, H. Yanagisawa, and Y. Miyoshi, Proceedings of the 7th International Conference on Nuclear Criticality Safety 2003 (2003).
4. M. Sasaki, T. Nakamura, N. Tsujimura, O. Ueda, and T. Suzuki, Nucl. Instr. and Meth. A 418 465–475 (1998).
5. A. Aroua and M. Hofert, Nucl. Instr. and Meth. A 372 318-321 (1996).
6. M. Luszik-Bhadra, Radiation Protection Dosimetry Vol. 96 Nos 1-3 pp. 227–229 (2001).
7. K. Tsuchiya, K. Kuroki, K. Kurosawa, N. Akiba, K. Tonoike, G. Uchiyama, Y. Miyoshi, H. Sono, T. Horita, K. Futakami, T. Matsumoto, J. Nishiyama,

and H. Harano, IEEE Nuclear Science Symposium 2010, Knoxville, USA Conference Records pp506–507 (2010).

8. A. Ochi, T. Nagayoshi, S. Koishi, T. Tanimori, T. Nagae, and M. Nakamura, Nucl. Instr. and Meth. A 478196–199 (2002).

9. K. Hattori, K. Tsuchiya, K. Ito, Y. Okada, K. Fujii, H. Kubo, K. Miuchi, M. Takada, T. Tanimori, and H. Uekusa, Journal of Synchrotron Radiation Vol. 16 Part 2 231–236 (2009).

10. J.D. Parker, K. Hattori, H. Fujioka, M. Harada, S. Iwaki, S. Kabuki, Y. Kishimoto, H. Kubo, S. Kurosawa, K. Miuchi, T. Nagae, H. Nishimura, T. Oku, T. Sawano, T. Shinohara, J. Suzuki, A. Takada, T. Tanimori, and K. Ueno, Nucl. Instr. and Meth. A 697 23 (2013).

Compact Short-Pulsed Electron Linac Based Neutron Sources for Precise Nuclear Material Analysis[*]

M. Uesaka[†], K. Tagi, D. Matsuyama, T. Fujiwara and K. Dobashi

Nuclear Professional School, University of Tokyo,
2-22 Shirane Shirakata, Tokai-mura, Naka-gun, Ibaraki 319-1188, Japan,
[†] *uesaka@nuclear.jp*
www.nuclear.jp/~kiki

M. Yamamoto

Accuthera Inc.,
2-8-22 Kurigi Asaoku Kawasaki-city, Kanagawa 215-0033, Japan
yamamoto@accuthera.com

H. Harada

Japan Atomic Energy Agency,
2-4 Shirane Shirakata, Tokai-mura, Naka-gun, Ibaraki 319-1195, Japan,
harada.hideo@jaea.go.jp

An X-band (11.424GHz) electron linac as a neutron source for nuclear data study for the melted fuel debris analysis and nuclear security in Fukushima is under development. Originally we developed the linac for Compton scattering X-ray source. Quantitative material analysis and forensics for nuclear security will start several years later after the safe settlement of the accident is established. For the purpose, we should now accumulate more precise nuclear data of U, Pu, etc., especially in epithermal (0.1-10 eV) neutrons. Therefore, we have decided to modify and install the linac in the core space of the experimental nuclear reactor "Yayoi" which is now under the decommission procedure. Due to the compactness of the X-band linac, an electron gun, accelerating tube and other components can be installed in a small space in the core. First we plan to perform the time-of-flight (TOF) transmission measurement for study of total cross sections of the nuclei for 0.1-10 eV energy neutrons. Therefore, if we adopt a TOF line of less than 10 m, the macro-pulse length of generated neutrons should be shorter than 100 ns. Electron energy, macro-pulse length, power, and neutron yield are ~30 MeV, 100 ns – 1 micros, ~0.4 kW, and ~10^{11} n/s (~10^3 n/cm^2/s at samples), respectively. Optimization of the design of a neutron target (Ta, W, ^{238}U), TOF line and neutron detector (Ce:LiCAF) of high sensitivity and fast response is underway. We are upgrading the electron gun and a buncher to realize higher current and beam power with a reasonable beam size in order to avoid damage of the neutron target. Although the neutron flux is limited in case of the

---

[*] This work is supported by etc, etc.
[†] Work partially supported by grant 2-4570.5 of the Swiss National Science Foundation.

X-band electron linac based source, we take advantage of its short pulse aspect and availability for nuclear data measurement with a short TOF system. First, we form a tentative configuration in the current experimental room for Compton scattering in 2014. Then, after the decommissioning has been finished, we move it to the "Yayoi" room and perform the operation and measurement.

Keywords: X-band electron linac based neutron source; decommsioned reactor; precise nuclear data study; fuel debris in Fukushima.

1. Introduction

We plan to use our X-band electron linac (11.424GHz, 30 MeV) [1] as a linac based neutron source [2, 3] for the nuclear data study for the Fukushima nuclear plant accident. Originally we developed the linac for Compton scattering X-ray source. Quantitative material analysis and forensics will start several years later after the safe settlement of the accident is established. For the purpose, we should now accumulate more precise nuclear data of U, Pu, etc., especially in epithermal (0.1-10 eV) neutrons. Therefore, we have decided to suspend the Compton scattering X-ray experiments and allocate the linac in the core space of the experimental nuclear reactor "Yayoi" which is now under the decommission procedure. Yayoi is the experimental fast neutron reactor (<1 MeV neutron, 2 kW, 10^{11} neutrons/ cm^2 /s). Especially, due to the compactness of the X-band 30 MeV linac, it can be installed into the fuel core space. Therefore, we can reuse the whole shielding structure of the reactor, the neutron beam-lines, and the control room. It is becoming more important to maintain the activity as a neutron source facility recently. This is because several old experimental reactors are going to be shut down.

2. 30 MeV X-band (11.424GHz) Electron Linac

As the RF frequency of electron linac is higher, its accelerating structure becomes smaller. Therefore, it can be even a mobile system. Recently, X-band linac (11.424 or 12 GHz) is adopted such a backup for international linear collider, which is CLIC by CERN, FEL by Elettra Syncrotrone Trieste, and Compton scattering γ-ray system for nuclear physics in ELI (Extreme Light Infrastructure) [4].

Our 30 MeV X-band (11.424 GHz) linac is basically designed and operated as a Compton scattering monochromatic X-ray source. The X-band linac consists of the 3 MeV thermionic RF gun, solenoid magnet for focusing, α magnet as an energy filter, 700 mm accelerating tube, and other components. A 50-MW X-band klystron and a 500-kV 1-μs modulator are used. ~10^4 micro-bunches of 20 pC and 1 ps(rms) forms 200 mA for 1 μs and 10 μA in average at 50 Hz. The macro-pulse length can be tuned down to 100 ns. The 3 MeV thermionic RF gun, solenoid magnet, and α magnet are adopted for a low emittance beam with the radius of 0.1 mm (rms) at the collision point with our

YAG laser. However, not low emittance but high average current is crucial as the neutron source. Therefore, the low energy part is replaced with 20 keV triode thermionic gun and 5 MeV traveling wave buncher. The electron gun can emit a beam of up to 400mA at 20kV. The capture efficiency of the buncher is about 1/2-1/3 and the buncher accelerate to 5MeV with 150mA. The regular accelerating structure accelerates electrons up to 30MeV. The average beam power is 375 W. The electron and X-ray beam spots are kept to be ~2 mmϕ in order to avoid the damage of the neutron target. Figure 1 shows a schematic drawing of the system. The important specifications are summarized in Table 1.

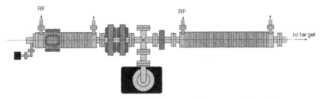

Fig. 1. Schematic drawing of the upgraded X-band 30 MeV linac specialized for neutron source.

Table 1 Specification of the upgraded X-band linac

Linac					
Electron beam energy	30	MeV	Group velocity/c	4(average)	%
Beam current (at target)	150	mA	Attenuation coefficient	0.5(average)	Naper/m
Pulse width	0.1-1 (variable)	μsec	Regular accelerator		
Repetition rate	50	pps	Number of cells	60	
Average beam power	0.375	kW	RF section length	0.523	m
Electron Gun			Q (Quality factor)$_0$	6593	
Voltage	20	kV	Shunt Impedance	95.6(average)	MΩ/m
Beam current	400 (max)	mA	Group velocity/c	3.97(average)	%
Buncher accelerator			Attenuation coefficient	0.478(average)	Naper/m
Number of cells	30		RF source		
RF section length	0.26	m	Pulse width	1	μsec
Q$_0$	6000		Power in buncher accelerator	6	MW
Shunt Impedance	85(average)	MΩ/m	Power in regular accelerator	35	MW

3. Design of Neutron Target, Moderator, Detector of TOF Line for Nuclear Data Analysis

3.1. *TOF analysis for more precise nuclear data of nuclear materials*

We should prepare more precise nuclear data of U, Pu and related nuclei for the quantitative material analysis and forensics of melted fuel and structural materials. A neutron active method is most promising for the purpose. First, we plan to perform the TOF transmission measurement for determining the total cross sections of the nuclei for 0.1-10 eV neutrons as shown in Fig.2. Uncertainty of the data of Pu in this region contains ~5% [5] while less than 1% for thermal neutrons. We have to upgrade the precision down to a few %

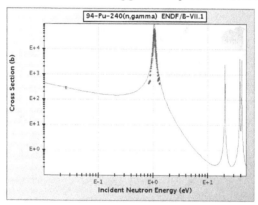

Fig. 2. (n, γ) Cross section of 94-Pu-240 for 0.1-10 eV neutrons.

One example of the time-energy relation of neutron obtained by J-PARC is shown in Fig. 3. The 0.1 – 10 eV energy range corresponds to ms time delay at TOF. In order to get this range and resolution at a TOF line of less than 10 m, the pulse length of electron and neutron should be around 100 ns.

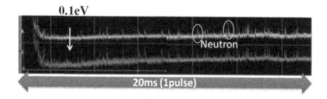

Fig. 3. Example of the time – energy relation of neutron obtained by J-PARC.

3.2. Neutron target and moderator

In case of an electron linac based neutron source, photon neutron reactions are used for generating neutrons by using a high atomic number material. Requirement for the target and moderator is to generate as many neutrons as possible and as a short pulse as 100 ns. This is because we want to perform nuclear data acquisition of nuclear materials for neutrons of energies of less than 100 keV at a 5 m TOF line. The Monte Carlo simulation code, the PHITS code,is used to optimize the target and moderator and to generate short pulsed neutrons with 30 MeV electrons [6]. We are trying several materials of W, Ta, Pb and U-238 for the target, light/heavy waters w/o Be for moderator. Typical geometry of the target and moderator is depicted in Fig.4. Typical neutron fluxes from the targets are shown in Fig. 5. Figure 6 shows typical pulse shapes of 10-100 keV for 10-MeV neutron incidence. Figure 7shows the energy spectrum of neutrons with the moderator of light water. Table 2 presents the target intensity, pulse width, and neutron flux at a 5 m TOF with maximum electron energy of 30 MeV and beam power of 375 W.

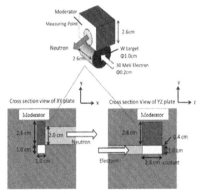

Fig. 4. Cross section view of the target and moderator.

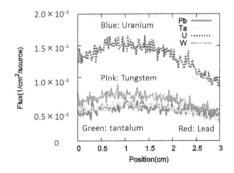

Fig. 5. Flux distribution with various target materials.

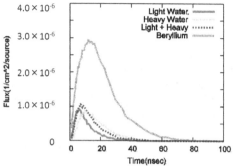

Fig. 6. Time resolution of 10 -100 keV neutrons with various moderator materials.

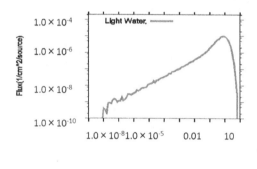

Energy(MeV)

Fig. 7. Energy spectrum of neutrons from moderator

Table 2 Neutron property at 5 m TOF with 30 MeV and 375 W electrons.

Target intensity	1.3×10^{11} n/s
Pulse width behind moderator (10 – 100 keV neutron)	6.66 ns
Neutron flux at measurement point (5m TOF)	1.1×10^{3} n/cm^{2}/s

3.3. Detector

In order to measure high precision nuclear data, the neutron detector must have a large neutron reaction cross section and fast decay time. ^{10}B or ^{3}He gaseous detector which has a large neutron cross section and low gamma-ray sensitivity, but due to the slow rise time of those detectors, they are not able to be used in high precision TOF measurement. On the other hand, inorganic neutron scintillators have a great characteristic in timing resolution. Especially, the scintillator called Ce:LiCAF, recently developed fast inorganic neutron scintillator, the light decay time is 40 ns which is superior to other neutron scintillators [7]. However, an electron linac based neutron source emits intense gamma-ray, and gamma-ray sensitivity of inorganic scintillators would be an issue for the precise neutron measurement.

There are two methods of discriminating between neutron and gamma-ray. One is a method using the difference of the range of the secondary particles. The other is a method which uses the difference of the light decay time of the scintillator.

When a Ce:LiCAF crystal is irradiated with a neutron, ^{6}Li absorb the neutron and emits charged particles, such as alpha-particle and triton. These secondary particles are absorbed in the crystal and they excite the crystal and emit photons. When it is irradiated with gamma-ray, the secondary particle would be electrons. The range of alpha-particle and triton are 10 μm orders of magnitude, but that of electron is 1 mm order [8]. Therefore, by controlling Ce:LiCAF crystal to suitable size that alpha-particle and triton will be absorbed but electron will be not, the discrimination ability will be improved. The single crystal rod and 10 mm× 10 mm × 2 mm sample of Ce:LiCAF are shown in Fig.8. Figure 8 shows the result of neutron/gamma-ray measurement by the pulse height analysis using a multi-channel analyzer, where a 2 mm × 2 mm × 2 mm crystal and a 10 mm× 10 mm × 2 mm crystal is irradiated by 100-keV neutrons from and 2.2-MeV gamma-rays from. It is clear that the discrimination for the 10 mm× 10 mm × 2 mm crystal is the best among them. Thus, the crystal would be adopted at first.

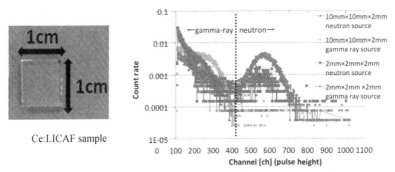

Ce:LICAF sample

Fig.8. Change of gamma-ray signal level following the square measure of crystal.

3.4. Two-phased development, operation and measurement

We take the two phased development, operation and measurement. First, we form a tentative configuration in the current experimental roomfor the Compton scattering in 2014. Then, after the decommissioning of Yayoi has been finished, we move it to the "Yayoi" room and perform the operation and measurement. The new 20-keV electron gun and 5-MeV buncher are replaced with the current 3-MeV thermionic RF gun. The new neutron target is installed just after the accelerating structure. Then, the samples and neutron detectors for nuclear data measurement are set so as to realize ~5 m long TOF line, which is almost equivalent to the 5 m long TOF line in the Yayoi room. Schematic drawing of allocation of the linac to the Yayoi area is shown in Fig. 9. The accelerating structure with the new 20-keV gun and 5-MeV buncher are inserted in the core space of the reactor. The klystron and modulator are put aside the shielding structure. 5, 10, 40 m long TOF lines are planned to be built step-by-step as shown in Fig. 9. The 5-m long TOF line can be set in the Yayoi room. Another experimental room exists next to the Yayoi room, which appears in the upper side in Fig. 9. The 10-m long TOF can be formed crossing across the wall between the two rooms via the beam port. The 40-m long TOF line is also available forward outside the building, which is depicted in the right-hand side in Fig. 9. We used to have and depose of this, but we plan to rebuild it with an advanced detecting system. We plan to start the neutron generation and experiment in early 2014. We shall start the nuclear data measurement with Au samples in the epithermal neutron energy region of 0.1-10 eV. This is because the nuclear data of Au in the neutron energy region poses sufficient precision and is good for calibrating the whole system. Then, we shall move the linac and 5-m long TOF line to the Yayoi room after the Yayoi decommission. Then, we plan to use the sample of nuclear materials of U, Pu, etc. there.

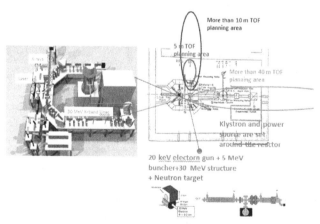

Fig. 9. Allocation of the linac to Yayoi area.

4. Future Nuclear Fuel Debris Analysis in Fukushima

Quantitative material analysis and forensics for nuclear security will start several years later after the safe settlement of the problems and decontamination are finished. A nuclear fuel debris analysis facility is now considered and designed by JAEA as shown in Fig. 10 (a) [9]. A ~1 kW class electron linac based neutron source is assumed to be used there. Both the neutron resonance transmission analysis (NRTA) and nuclear resonance capture analysis (NRCA) are used. Typical numerical results of NRTA are given in Fig.10 (b) [10, 11].

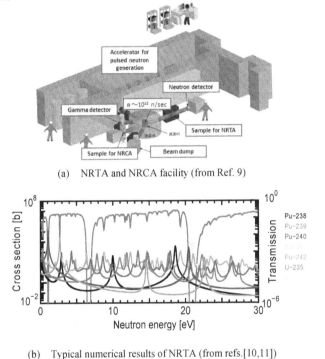

(a)   NRTA and NRCA facility (from Ref. 9)

(b)   Typical numerical results of NRTA (from refs.[10,11])

Fig. 10. Expected Neutron Transmission Result by Using the Precise Nuclear Data.

5. Conclusion

An X-band (11.424GHz) 30-MeVelectron linac as a neutron source (~10^{11} n/s, ~10^3 n/cm^2/s at samples) for the nuclear melted fuel analysis and nuclear security in Fukushima is developed. We are manufacturing a 20-keV triode electron gun and 5 MeV traveling wave buncher to realize higher current and beam power (~0.4 kW) with a reasonable beam size in order to avoid damage of the neutron target. Optimization of the design of the neutron target, a TOF line and a neutron detector (Ce:LiCAF) of high sensitivity and fast response is underway. We plan to perform the two-phased development and measurement. Although the neutron flux is limited in case of the X-band electron linac based source, we take advantage of its short pulse aspect and availability for nuclear data measurement with a short TOF system. First, we form the preliminary configuration in the current experimental room in 2014. Then, after the decommission has been finished, we move it to the "Yayoi" room and perform the operation and measurement.

References

1. F. Sakamoto, M. Uesaka, Y. Taniguchi, T. Natsui, E. Hashimoto, L.K. Woo, T. Yamamoto, J. Urakawa, M. Yoshida, T. Higo, S. Fukuda, N. Kaneko, H. Nose, H. Sakae, N. Nakamura, M. Yamamoto, "Compton sources for X/γ rays: Physics and applications", Nuclear Instruments and Methods in Physics Research Section A: Accelerators, vol. 608, S36-S40(2009)
2. M.S. de Jong, "PRODUCING MEDICAL ISOTOPES USING X-RAYS", THXA01, Proc. of IPAC2012 (New Orleans)
3. H. Kobayashi, et al.,," CONSTRUCTION OF A BNCT FACILITY USING AN 8-MeV HIGH POWER PROTON LINAC IN TOKAI", THPPR048, Proc. of IPAC2012 (New Orleans)
4. http://indico.cern.ch/conferenceDisplay.py?confId=231116
5. R.R. Spencer et al., Nucl. Sci. Eng. 96, 318-329; 1987.
6. K. Niita, T. Sato, H. Iwase, H. Nose, H. Nakashima, and L. Sihver, "PHITS—a particle and heavy ion transport code system," *Radiat. Meas.*, vol. 41, no. 9, pp. 1080–1090, 2006.
7. T. Fujiwara, H. Takahashi, T. Yanagida, et al., "Study on Ce:LiCAF scintillator for ^{3}He alternative detector", Neutron News Volume23 Number 4, p.31-34 (2012)
8. G. F. Knoll, "Radiation Detection and Measurement THIRD EDITION", trans. Itsuro Kimura and Eiji Sakai, 2001
9. M. Koizumi et al., Proc. of INMM 53th annual meeting; 2012.
10. H. Harada et al., "Neutron resonance densitometry for particle-like debris of melted fuel, to be published in Nuclear Data Sheets
11. H. Harada et al., 2013 Annual Meeting of the Atomic Energy Society of Japan; 2013; A54.

Laser-Driven Plasma Deceleration of Electron Beams for Compact Photon Sources

J.-L. Vay*, C. G. R. Geddes, S. G. Rykovanov, C. B. Schroeder,

E. Esarey and W. P. Leemans

Lawrence Berkeley National Laboratory,
Berkeley, CA 94720, USA
jlvay@lbl.gov

Particle accelerators are in widespread use as intense, precisely controllable photon sources, but many applications, including nuclear nonproliferation, are limited by size. Laser-driven plasma accelerators (LPAs) reduce accelerator size, but a compact system also requires addressing radiation hazards resulting from disposal of particle beam energy after photon production, typically requiring large and heavy "beam dumps". In this paper, we investigate, through 3-D Particle-In-Cell simulations, an LPA stage demonstrating equal effectiveness at accelerating and decelerating an electron beam over a very short distance. This indicates that in addition to providing compact accelerators, such structures can effectively reduce beam energy after photon production and hence beam dump weight and volume. This is important to the development of compact photon source systems which can satisfy needs including transportable operation or operation in populated areas.

Keywords: Particle accelerators; plasma-based acceleration; laser-driven plasma acceleration; beam dump.

1. Introduction

Particle accelerators delivering high-quality charged particle beams are in widespread use as photon sources using coherent or incoherent intrabeam particle oscillations induced with e.g., free-electron lasing, betatron oscillations or Thomson scattering of a counter-propagating laser (also referred to as Compton or Inverse Compton scattering). Disposal of the high energy particle beam after photon production however imposes the use of large and heavy "beam dumps" that usually prevent portability of the photon source, limiting applicability. For example, transportable sources of quasi-monoenergetic MeV photon beams would offer unprecedented versatility for nonproliferation applications including active interrogation of cargo containers to search of Special Nuclear Material (SNM)[1,2], treaty verification and Non-Destructive Assay (NDA) of spent nuclear fuel[3]. Thomson scattering photon sources are a leading candidate to meet these needs due to their narrow energy spread and divergence, and tunable energy, but typically require 0.2-0.6 GeV electron beams[2,4]. For portability, constraints of size and weight require acceleration of the electron beam in a short distance and also disposal of its energy (after photon production) in a way that minimizes size and the use of heavy materials. High beam currents are also required to meet application needs for photon flux, compounding the problem. Accelerators based on standard technology do not meet these constraints, and novel acceleration methods are needed.

Laser-driven plasma based electron accelerators (LPAs)[5,6] have been demonstrated to produce high-quality electron beams at the required energies in cm-scale distances[7,8], fulfilling the need for compact acceleration. With such small accelerators, the size of a photon source would be dominated by conventional methods for disposal of the electron beam which require heavy shielding. For the beam energies required to produce MeV photons, this shielding is of room size and can preclude transportable operation. In principle, the same structure used for acceleration in an LPA can be used to decelerate the electron beam by appropriately phasing the beam in the plasma wake. This can theoretically decelerate the beam over the same cm-scale distance as required for acceleration. Limitations include dephasing with regard to deceleration and focusing as well as energy spread from non-uniform deceleration.

In this paper, we investigate through 3-D Particle-In-Cell simulations the use of an LPA for the active deceleration of the electron beam after photon production.

Photon sources based on LPAs for nonproliferation applications have been studied elsewhere[9], showing that for many applications photon production within the accelerating plasma produces sufficient beam quality and that the electron beam distribution is not greatly affected by the photon production process. Hence for simplicity we focus here on such a single plasma structure, and characterize the acceleration and deceleration phases. Dependence of deceleration efficacy on plasma length and phasing, including self consistent laser evolution, have been characterized to maximize deceleration.

2. Principles of laser-driven plasma deceleration of electron beams

In laser-driven plasma accelerators, a laser displaces electrons in a plasma channel, initiating plasma oscillations and resulting in a succession of positively and negatively electrically charged regions behind the laser (or "wake"). The alternating polarity within the wake generates very strong (typically GV/m) longitudinal and transverse electric fields of alternating sign. An electron beam located at the appropriate phase behind the laser will be both focused transversely and accelerated to (or decelerated from) high energies over a very short distance. The accelerating and focusing fields are driven by the ponderomotive force of the laser pulse $F \simeq -m_e c^2 \nabla a^2/2$, where $a = eA/m_e c^2$ is the normalized amplitude of the laser vector potential A, e and m_e are the electron charge and mass respectively, and c is the speed of light in vacuum[6]. The longitudinal field is of the order of $E_p(\text{V/m}) \simeq 96\sqrt{n_0(\text{cm}^{-3})}$ with n_0 the plasma electron density, which can reach several orders magnitude higher amplitudes than with conventional acceleration techniques.

Fig. 1 shows the transverse and longitudinal electric fields in the wake produced by a laser in the quasi-linear regime ($a_0 = 1$). The alternating focusing-defocusing and accelerating-decelerating periods are shifted by $\sim \pi/2$, and approximately half a period is available for focusing. This focusing phase is then split between accelerating or decelerating regions for the electron beam. The group velocity of the laser in the plasma is typically smaller than that of the electron beam, such that slippage occurs. Hence, an electron beam injected appropriately will be focused and accelerated to very high energy in a very short distance, then will slip ahead in the wake and reach the phase where it is efficiently decelerated while still being focused. This phasing effect has been studied in depth to enable extraction of the electron beam from the LPA at peak energy[6,8,10–12]. By continuing the plasma beyond the length at which the beam achieves peak energy, acceleration and deceleration are accomplished in the same compact (cm-scale) structure. For moderate energy spreads, photon production can then be conducted in the same plasma, at the phase interval between accelerating and decelerating regions where the longitudinal field is near zero[9]. For very low energy spreads manipulation of beam divergence or phasing may be needed, but the principle remains the same[9].

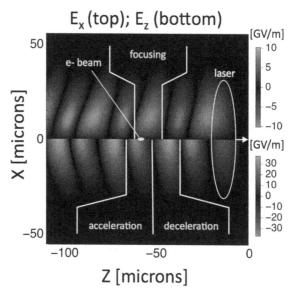

Fig. 1. Snapshot from 2-D PIC simulation of laser-driven plasma accelerator showing the wakefield from a laser propagating from left to right in a plasma channel: (top) transverse focusing (red) and defocusing (blue) fields; (bottom) accelerating (blue) and decelerating (red) fields. With appropriate phasing of the injection of the electron beam (green), approximately half of a period is available for simultaneous guiding and acceleration, immediately followed by simultaneous guiding and deceleration of the electron beam.

3. Results from Particle-In-Cell simulation of a single acceleration and deceleration stage

This section presents results from a 3-D simulation of a single laser-driven accelerating and decelerating plasma stage using the Particle-In-Cell code Warp[13]. Simulating electron accelerators that use laser-driven plasma waves (wakefields) can be challenging due to the disparity in scale between the micron-scale laser wavelength and the acceleration length, as well as the large number of plasma particles. Recasting the simulation in a Lorentz boosted frame moving in the direction of the wake enables use of very high resolution, high concurrency simulations at dramatically reduced wall clock time with effectively no loss of relevant physics in the modeling of laser plasma accelerators[14]. The simulation that is reported here was run using a Lorentz boosted frame with a boost at $\gamma_{boost} \sim 16.6$ for a speedup[15] near $4\gamma_{boost}^2 \sim 1100$, which allowed rapid turnaround for parametric exploration in three-dimensions at full scale.

For production of 6 MeV photons from Thomson scattering, the desired energy of the beam after acceleration was 0.5 GeV and relative energy spread at or below 2%[9]. The parameters of the simulation to achieve this were determined from previous simulations using scaling laws that have been demonstrated over a very wide range of energies[15]. These scaling laws allow predictive design of LPA stages over a wide range of energies, and show that parameters such as energy spread remain constant. A laser of wavelength $\lambda = 0.8\mu m$, profile $a(r, z) = a_0 \exp\left[-r^2/w^2\right] \sin\left[\pi z/L\right]$ with a waist $w \sim 25\mu$m, length $L \sim 28\mu$m and amplitude $a_0 = 1$, was injected in a plasma column of density $n_0 = 1.3 \times 10^{18}$ cm^{-3} on axis with a parabolic transverse profile which provides laser guiding. An electron beam was injected in the wake with charge $Q = -10$pC, energy $E \sim 27$ MeV and relative energy spread $\Delta E/E \sim 0.1$, a Gaussian profile with R.M.S. width $\sigma_x = \sigma_y \sim 0.56\mu$m and length $\sigma_z \sim 0.5\mu$m and a normalized R.M.S. emittance $\epsilon_x = \epsilon_y \sim 33\mu$m.mrad. These are consistent with LPA injector parameters measured experimentally and reported elsewhere[16]. The beam was injected into the second plasma oscillation at the phase for acceleration and guiding. This phase corresponded to $D \sim 50\mu$m behind the peak of the laser pulse, or $D \sim 1.7\lambda_p$ (where $\lambda_p \sim 2\pi c/\omega_p$ is the wake wavelength, $\omega_p = \sqrt{(n_e e^2)/(\epsilon_0 m_e)}$ is the plasma frequency with n_e the plasma density and ϵ_0 the permittivity of vacuum).

Fig. 2 shows the plasma profile on axis, electron beam average energy and energy spread histories, as well as snapshots of the transverse and longitudinal electric fields and electron beams. The snapshots are taken from a simulation using a Lorentz boosted frame (boost at $\gamma_{boost} \sim 16.6$) and Z', E'_x and E'_z are respectively the longitudinal coordinate, transverse electric field and longitudinal electric field in the simulation frame. The electron beam is accelerated to 0.5 GeV in 2 cm, then decelerated to its injected energy in a slightly shorter distance of 1.5 cm, the asymmetry of the acceleration and deceleration distances being attributed to steepening of the wake structure from laser depletion[17]. The relative energy spread falls from 10% at injection to slightly below 2% at peak energy (as prescribed for Thomson scattering), then rises steadily to nearly 100% at the plasma exit at $z = 34$mm. The beam energy is within 2 % of the peak value (0.5GeV) over the central 2 mm of propagation. The length of this region is similar to the scattering pulse length required, as shown elsewhere[9]. Using the electron beam parameters from the Warp simulation, a simulation was performed with VDSR[4] to compute the spectrum of photons from Thomson scattering giving $\approx 13\%$ FWHM photon energy spread at 6 MeV,

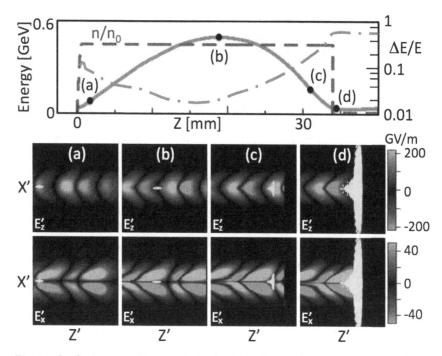

Fig. 2. (top) plasma profile on axis (dash blue), electron beam average energy (solid red) and energy spread histories (dot-dash green) in the laboratory frame; (bottom) snapshots of the transverse and longitudinal electric fields and electron beams (green) taken at propagation distances (a), (b) (c) and (d) indicated on top plot, in the Lorentz boosted simulation frame.

assuming a counter-propagating laser of wavelength $\lambda = 0.8$ μm, amplitude $a_0 = 0.05$, length 27λ and focus width of 20λ. It has been previously demonstrated that the LPA energy and beam performance scale predictably with plasma density[15]. This allows other photon energies (e.g. 1 to 15 MeV) to be achieved with similar beam parameters to the present example. Snapshots of the fields and electron beam at various times show that the beam is well focused when it enters the decelerating region of the wake but reaches the defocusing region toward the end of its deceleration, causing the beam to spread significantly transversely. The transverse spread induces non-uniform deceleration that results in higher energy spread and limits deceleration efficiency. Through parametric exploration that was enabled by the fast turnaround of Lorentz boosted frame simulations, the plasma length was adjusted to ~ 34mm, which maximized averaged deceleration and minimized energy spread for the chosen laser, plasma and electron beam injection phase parameters.

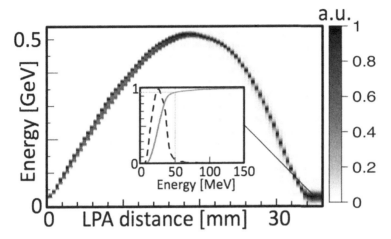

Fig. 3. Energy profile of the electron beam versus propagated distance in the plasma channel (in the laboratory frame); (insert) beam energy distribution (black dash) and cumulative energy distribution (solid red at exit).

Fig. 3 shows the energy profile of the electron beam versus propagated distance, as well as the beam energy distribution at exit, demonstrating efficient deceleration with 95% of the electron beam energy below 50 MeV. This constitutes deceleration to less than 10% of the maximum energy .

This study demonstrates the potential for efficient acceleration and deceleration over very short distances with LPAs. With regards to deceleration, further improvements are possible through the use of e.g., longitudinal plasma shaping and the addition of a passive deceleration plasma region.

4. Conclusion

It has been demonstrated that an LPA can efficiently accelerate an electron beam to 0.5 GeV with a relative energy spread around 2% and efficiently decelerate this same beam with 95% of the electrons below 50 MeV. Because the beam is decelerated by collective fields rather than by scattering, high energy radiation is not produced during deceleration. Hence, the beam dump that is needed is similar to that for a < 50 MeV electron beam. This study provides an example demonstrating the effectiveness the laser-driven plasma deceleration of electron beams. Advanced photon sources may incorporate multiple plasmas to reduce bandwidth[9] which would allow greater freedom in controlling deceleration, and hence more effective deceleration than shown here. These results are important for application to the development of portable photon sources, such as sources of quasi-monoenergetic MeV photon beams for nuclear nonproliferation applications. The narrow energy spread at peak energy is suitable for efficient X-ray production through Thomson scattering with a counter propagating laser, while the efficient reduction of electron beam energy will enable compact portable disposal and background/operator dose that are reduced to acceptable level for operation in populated areas. Further work will incorporate the modeling of the scattering region for very high photon flux cases where scattering may perturb the beam, and explore further deceleration via longitudinal plasma shaping (which can mitigate defocusing) and the addition of a passive deceleration plasma region.

Acknowledgments

This work was supported by US-DOE Contracts DE-AC02-05CH11231 and DE-AC52-07NA27344, by US DOE NNSA DNN R&D, and US-DOE Sci-DAC program ComPASS, and used resources of NERSC, supported by US-DOE Contract DE-AC02- 05CH11231.

References

1. B. J. Quiter, et al., *J. Applied Phys.* **103** (2008).
2. J. Pruet, et al., *J. Applied Phys.* **99** (2006).
3. R. Hajima, et al., *Nucl. Inst. Meth. in Phys. Res. A* **608** 2009)
4. M. Chen, et al., *Phys. Rev. STAB* **16** (2013).
5. T. Tajima and J. Dawson, *Phys. Rev. Lett.* **43**, 267 (1979).
6. E. Esarey, et al., *Rev. Mod. Phys.* **81**, 1229 (2009).
7. N. H. Matlis, et al., *J. Opt. Soc. America B-Opt. Phys.* **28**, 23 (2011).
8. W. P. Leemans, et al., *Nature Phys.* **2**, 696 (2006).
9. S. G. Rykovanov, *in preparation* .
10. C. Geddes, et al., *Nature* **431**, 538 (2004).
11. S. Mangles, et al., *Nature* **431**, 535 (2004).
12. J. Faure, et al., *Nature* **431**, 541 (2004).
13. J.-L. Vay, et al., *Comput. Sci. and Discovery* **5**, 014019 (2012).
14. J.-L. Vay, et al., *Phys. Rev. Lett.* **98**, 130405/1 (2007).
15. J. L. Vay, et al., *Phys. Plasmas* **18** (2011).
16. A. J. Gonsalves, et al., *Nature Phys.* **7**, 862 (2011).
17. B. A. Shadwick, et al., *Phys. Plasmas* **16**, 056704 (2009).

Advanced Laser-Compton Gamma-Ray Sources for Nuclear Materials Detection, Assay and Imaging

C. P. J. Barty

Lawrence Livermore National Laboratory,
Livermore, CA 94551 USA
barty1@llnl.gov

Highly-collimated, polarized, mono-energetic beams of tunable gamma-rays may be created via the optimized Compton scattering of pulsed lasers off of ultra-bright, relativistic electron beams. Above 2 MeV, the peak brilliance of such sources can exceed that of the world's largest synchrotrons by more than 15 orders of magnitude and can enable for the first time the efficient pursuit of nuclear science and applications with photon beams, i.e. Nuclear Photonics. Potential applications are numerous and include isotope-specific nuclear materials management, element-specific medical radiography and radiology, non-destructive, isotope-specific, material assay and imaging, precision spectroscopy of nuclear resonances and photon-induced fission. This review covers activities at the Lawrence Livermore National Laboratory related to the design and optimization of mono-energetic, laser-Compton gamma-ray systems and introduces isotope-specific nuclear materials detection and assay applications enabled by them.

Keywords: Gamma-rays; nuclear resonance fluorescence; laser-Compton scattering.

1. Introduction

For more than a decade, the Lawrence Livermore National Laboratory (LLNL) has been developing and advancing the state of the art with respect to laser-based Compton x-ray and gamma-ray light sources. In recent years, work has concentrated on the optimization of laser-Compton gamma-ray sources with respect to size, flux and bandwidth with aspecific aim of creating field-worthy, mono-energetic gamma-ray systems for materials detection, assay and imaging applications. Mono-energetic gamma-rays can efficiently excite the electromagnetic resonances of the nucleus and produce nuclear resonance fluorescence (NRF). NRF transitions depend upon the number of protons and neutrons in the nucleus and are thus unique fingerprints of each isotope. By monitoring the attenuation of gamma-ray photons due to the excitation of NRF transitions, one may determine the presence, amount and distribution of specific isotopes within arbitrary objects. Additionally, NRF transitions for actinides of interest to the nuclear power and nuclear security communities occur at photon energies that are highly penetrating, e.g. in the 1 MeV and 4 MeV spectral range. It is therefore possible to not only assay and detect materials with laser-Compton gamma-ray sources but to do so in the presence of appreciable shielding or inside of thick objects. Possible applications enabled by next generation laser-Compton gamma-ray sources include rapid (milliseconds) detection of concealed nuclear material, high precision (better than 100 parts per million) non-destructive assay of spent nuclear fuel assemblies, isotope-specific, high-resolution (less than 10 micron spatial resolution) 3D imaging of nuclear materials in existing waste containers and waste processing streams.

2. Laser-Compton Scattering

Compton scattering of laser photons from relativistic electrons was first demonstrated in 1965.[1] In that experiment a giant pulse ruby laser interacted with 6 GeV electrons and created approximately 8 upshifted photons per laser pulse. In the years following this demonstration, laser Compton scattering was used as a diagnostic of electron beam quality in advanced accelerators. In its simplest configuration, laser light is incident head on with the electron beam and the on axis, upshifted photons have an energy equal to $4\gamma2E_l$, where γ is the normalized energy of the electron and E_l is the incident photon energy. By monitoring the spectrum of the upshifted photons, one may learn about the energy spread the electron beam. In the 1990's a renaissance in laser-Compton scattering arose from the ultrafast materials community which used the process to produce short duration bursts of x-rays, typically of a few 100 fs to few ps in duration. [2,3] In order to reduce the duration of the resulting x-ray pulse in these systems, the laser was often incident at right angles to the electron beam direction. While these sources produced short duration x-rays, they also produced relatively broadband x-rays (>10% $\Delta E/E$), were relatively inefficient and the up-scattered photon energy was only half of that from a head on collision. Fundamentally the efficiency of laser Compton scattering is limited by the small magnitude of the Thomson cross section (~0.6 barns) and the inability of electron beams to be focused to spots on par with minimum laser spot dimensions. In 2004, LLNL scientists recognized [4] that the Compton scattering brilliance should increase rapidly as a function of electron beam energy and beam quality. To first order this occurs because at higher electron beam energy it is possible to overcome electrostatic repulsion and focus the electron bunch to smaller spot dimensions. Roughly, the electron spot dimension is proportional to its beam energy and thus the peak brilliance (photons/sec/0.1%BW/mrad2/mm2) of the laser-Compton source increases as function of electron beam energy somewhere between 2nd and 4th power. This rapid increase in peak brilliance is illustrated in Figure 1 and is in stark contrast to the trends of alternative sources, such as large-scale synchrotrons. In the nuclear excitation region above 100 keV, the peak brilliance of 3rd generation synchrotrons decreases faster than exponentially.

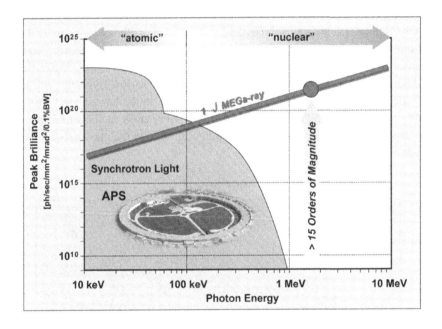

Figure 1. Peak brilliance of a laser-Compton light source relative to the APS synchrotron.

Above 2 MeV, the peak brilliance of an optimally-configured, laser-Compton source can exceed that of the largest synchrotrons by more than 15 orders of magnitude. It is important to note that for many nuclear applications and especially for those related to nuclear materials management and detection, it is the bandwidth of the Compton source and not the pulse duration or brilliance that is of foremost importance.

The optimization of laser-Compton scattering to produce narrowband gamma-rays involves a different approach to machine design than that pursued for short duration x-ray sources. The bandwidth of the Compton source is driven by three effects; the energy spread of the electron bunch, the bandwidth of the laser photons and the spread due to the angle correlation in the interaction region. Bandwidth can be minimized with high quality (low emittance) electron beams, few-ps or longer laser pulses and optimized laser-electron interaction geometries. Fractional bandwidths of ~10^{-3} $\Delta E/E$ or ~2 orders of magnitude less than that demonstrated from short-duration, laser-Compton x-ray sources are possible with careful design.

3. Excitation of Nuclear Resonance Fluorescence

Laser-Compton gamma-ray sources can enable "Nuclear Photonics", that is the photon-based manipulation and study of the nucleus. In particular narrow-band sources can efficiently excite nuclear resonance fluorescence (NRF). While the width of NRF transitions at room temperature is very narrow (typically 10^{-5} to 10^{-6} $\Delta E/E$), selective excitation is possible with an optimized, 10^{-3} $\Delta E/E$ bandwidth laser-Compton source. Furthermore, NRF cross sections of interest are large compared to background and often occur within the max-transparency window for most materials (see Figure 2).

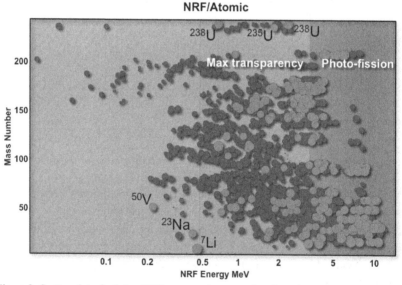

Figure 2. Scatter plot of relative NRF cross sections as a function of mass number and photon energy. Dot size is proportional to cross section magnitude. Red dots are larger than the atomic background at the particular energy.

Laser-Compton excitation of NRF can be used in "reflection" or in "transmission" to determine the presence or absence of a particular isotope. In reflection, one illuminates the object in question and looks for the characteristic NRF relaxation radiation which is emitted into 4π. In transmission one looks for the absence of resonant photons in the transmitted beam. Besides being intrinsically less susceptible to clandestine attempts to obscure signals, transmission based systems can also provide quantitative assay and high resolution spatial information regarding the isotopic content of the object. In transmission the primary issue is low angle Compton scattering which can create new photons at the resonance energy and degrade measurement accuracy. It has been numerically shown that this problem is alleviated for sufficiently narrowband and collimated laser-Compton sources. [5]

As an example let us consider a standard nuclear fuel rod containing isotopic defects which include variations in density and enrichment (see Figure 3). [6]

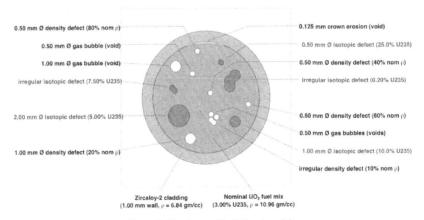

Figure 3. Uranium oxide fuel rod model.

As illustrated in Figure 4, the Bremsstrahlung image can identify the density defects but misses the enrichment variations. However an image obtained with [235]U-resonant, 1733-keV gamma-rays is able to identify the location and magnitude of the enrichment variations. Separate analysis [6] suggests that high-flux, narrowband, laser-Compton systems currently being constructed will be able to assay nuclear fuel assemblies with better than 100 ppm accuracy per isotope of interest.

Figure 4. Simulated Uranium oxide fuel rod images with 2 MeV Bremsstrahlung (left) and 1733 keV laser-Compton gamma-rays (right). The simulations were performed with a modified version of LLNL's Monte Carlo code – COG that includes NRF transitions and their line shapes.

4. Detector Systems for Laser-Compton Sources

Transmission based detection and assay of materials with narrow-band, laser-Compton gamma-rays is fundamentally an issue of observing the narrow-band (few eV wide) removal of resonant photons from the much wider (~1000 eV wide) spectrum of the interrogating laser-Compton beam. The detection and measurement of this "notch" in the transmitted spectrum is well beyond the ~1000 eV resolution of today's best single photon counting, gamma-ray spectroscopic techniques. Furthermore the high flux per pulse characteristics of laser-Compton sources is ill-matched to the single-photon-counting operation of traditional high-resolution gamma-ray detectors, e.g. high purity germanium (HPGe). To overcome these issues, a new, calorimetric, dual-isotope notch observation (DINO) detector arrangement was developed and patented (US8369480 B2) by LLNL. Schematically DINO detector arrangement is shown in Figure 5.

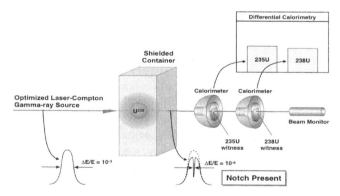

Figure 5. Schematic of the Dual Isotope Notch Observation (DINO) detector arrangement.

In the DINO arrangement illustrated in Fig. 5, the beam is transmitted through the interrogated object and interacts with two "witness" samples before being collected by a beam monitor. The witness samples are composed of two isotopes of the element that is sought in the interrogated object, e.g. ^{238}U and ^{238}U if interrogating a spent fuel rod. The first witness is composed of the isotope for which the laser-Compton beam may contain resonant photons. When the transmitted beam interacts with this witness Delbruck, Rayleigh, Thomson, Compton and NRF scattering will take place. On the other hand illumination of the second witness will only produce Delbruck, Rayleigh, Thomson and Compton scattering. Because the magnitudes of the non-resonant scattering cross sections (Delbruck, Rayleigh, Thomson and Compton) do not change rapidly with nucleon number, the scattering from the second witness provides an accurate measure of the non-resonant scattering present in the first witness sample. Thus the properly normalized difference of the integrated scattered energy from both samples is proportional to only the amount of NRF scattered by the resonant witness. The more resonant material that is present in the object, the more resonant photons that will be removed from the transmitted laser-Compton beam and the less difference in the normalized scattered energy from the two witness materials there will be. By placing known quantities of the desired isotopic material in the beam, it is possible to create an absolute calibration of the calorimetric difference between the scattered radiation from the two witnesses and thus to determine not only the presence of material in the interrogated object but also the precise amount of material in the object. Unlike reflection based detection systems, the DINO arrangement can tell by the lack of sufficient photons on the beam monitor if the object has been two heavily shielded to determine the presence of the desired isotopic material. This is an important attribute for nuclear materials security applications. Furthermore the DINO detector arrangement does not require the use of single photon counting detectors and thus can take full advantage of the high-photon-number per pulse characteristics of optimized laser-Compton sources.

The design metric of importance for NRF-based interrogation with laser-Compton sources is the ratio of the number of photons from the source that are resonant with the desired material per unit time (i.e. the signal) to the number of photons that are not resonant (i.e. the noise). This is proportional to the specific spectral density (SSD) of the source or the number of photons per second per unit bandwidth divided by the fractional bandwidth of the source. As illustrated in Figure 6, new detection, assay, imaging and science capabilities are enabled as the SSD of the source increases.

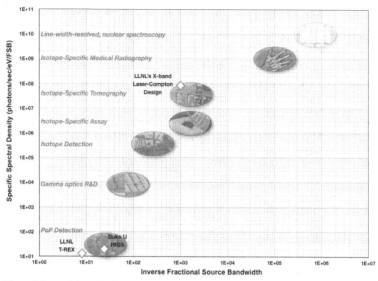

Figure 6. Potential applications enabled by increased source specific spectral density (SSD).

5. Optimized Laser-Compton Gamma-ray Systems

As illustrated in Figure 6, existing laser-Compton-based materials detection studies in the US (LLNL and Duke/HIγS) and Japan (AIST/JAEA) have utilized sources with SSD's in the range of 10 to 100 depending upon gamma-ray energy [7-9]. However, for practical detection and assay, a source SSD in the range of 1,000,000 to 10,000,000 is required. LLNL's approach to meeting these requirements is based uponhigh-gradient, compact, x-band accelerator technology first pioneered by the SLAC National Accelerator Laboratory. In order to produce the required beam current (micro-amps), the accelerator must be operated in a macro/micro bunch mode thatproduces, high-quality, low-emittance (0.1 mm-mrad-scale), electron beams[10] at repetition rates up to ~100 kHz. To test and refine this mode of operation a sub-scale x-band test system has been constructed and is being commissioned (see Figure 7) at LLNL.

The optimized version of this system utilizes a new, mulit-GHz, fiber-laser technology [11] to produce the sub-ps, few micro-Joule UV pulses for illumination of the photo-cathode and to produce the low-energy, chirped, IR pulses that seed the high-energy (> 1J), diode-pumped, 120-Hz, Nd:YAG laser-amplifiers that in turn produce the energeticlaser pulses used in Compton gamma-ray generation. The pulses from the fiber laser system are produced viaelectro-optic modulation of a CW laser at the same x-band (11.424 GHz) RF frequency that powers the accelerator. The resulting 11.424 GHz train of long-duration (~50 ps), low-energy (~pJ),laser pulses are then amplified in large-core fiber amplifiers to increase their energy and passed through lengths of small-core, passive fiber to increase their bandwidth via self phase modulation. The resulting micro-Joule scale, chirped pulses are then passed though a dispersive optical delay line (a parallel grating pair) and compressed to sub-ps duration. The significant advantage of this architecture is that it produces pulses that are intrinsically matched and synchronized to the x-band RF (11.424 GHz) frequency of the accelerator. This simplifies the machine and enables perfect overlap between individual electron bunches and the energetic laser pulses.

Figure 7. LLNL's compact, x-band accelerator R&D system. Shown are the high brightness photo-gun (below the researcher's hands) and the gun solenoid (left) and x-band accelerator section (right).

The effective average power of the energetic interaction laser can be significantly increased by pulse capture and recirculation through the interaction region. LLNL has previously demonstrated ~40x recirculation of energetic (Joule-class) laser pulses via injection of IR pulses into a dichroic, ring laser cavity containing a frequency conversion crystal [12]. Detailed modeling of the laser-Compton process based on codes that have been benchmarked against previous laser-Compton experiments at LLNL predicts that the combination of the above technologies and of optimized, head-on Compton scattering geometries will result in specific spectral densities that are 5 to 6 orders of magnitude beyond the present state of the art.

6. Conclusion

Next generation laser-Compton gamma-ray sources will enable the practical, isotope-specific, detection, assay and imaging of nuclear materials in a non-destructive and non-activating manner. In addition, the high-gradient compact accelerator technology and efficient, diode-pumped laser technologies that are being pursued at LLNL will also enable creation of robust, mobile, re-locatable gamma-ray systems. These technologies have the potential to significantly alter the methods and procedures by which all nuclear materials are managed and will fundamentally alter the way the photo-nuclear processes are studied.

Acknowledgment

This work was performed under the auspices of the U.S. Department of Energy by the Lawrence Livermore National Laboratory under Contract DE-AC52-07NA27344.

References

1. C. Bemporad, R. Milburn, N. Tanaka and M. Fotino, Phys. Rev. 138, 6B, (1965).
2. R. Schoenlein, W. Leemans, A. Chin, P. Volfbeyn, T. Glover, P. Balling, M. Zolotorev, M. Kim, S. Chattopadhyay, and C. Shank, Science 274, 236 (1996).
3. D. J. Gibson, S. G. Anderson, C. P. J. Barty, S. M. Betts, R. Booth, W. J. Brown, J. K. Crane, R. R. Cross, D. N. Fittinghoff, F. V. Hartemann, J. Kuba, G. P. Lesage, D. R. Slaughter, A. M. Tremaine, A. J. Wooton, E. P. Hartouni, P. T. Springer, and J. B. Rosenzweig, Phys. Plasmas 11, 2857 (2004).
4. C. P. J. Barty and F. V. Hartemann, "T-REX: Thomson-Radiated Extreme X-rays Moving X-ray Science into the "Nuclear" Applications Space with Thompson Scattered Photons", LLNL Technical Report, UCRL-TR-206825 (2004)
5. J. Pruet, D. P. McNabb, C. A. Hagmann, F. V. Hartemann, and C. P. J. Barty, J. Appl. Phys. 99, 123102 (2006).

6. J. M. Hall, V. A. Semenov, F. Albert and C. P. J. Barty, "Numerical simulation of nuclear materials detection, imaging and assay with MEGa-rays", Proc. 52nd Annual INMM Meeting (Palm Desert, CA, 2011).

7. F. Albert, S. G. Anderson, G. A. Anderson, S. M. Betts, D. G. Gibson, C. A. Hagmann, J. M. Hall, M. S. Johnson, M. J. Messerly, V. A. Semenov, M. Y. Shverdin, A. M. Tremaine, F. V. Hartemann, C. W. Siders, D. P. McNabb and C. P. J. Barty, Opt. Lett, 35, 3 354 (2010).

8. C. A. Hagmann, J. M. Hall, M. S. Johnson, D. P. McNabb, J. H. Kelley, C. Huibregtse, E. Kwan, G. Rusev, and A. P. Tonchev, J. Appl. Phys. 106, 084901 (2009).

9. N. Kikuzawa, R. Hajima, N. Nishimori, E. Minehara, T. Hayakawa, T. Shizuma, H. Toyokawa, and H. Ohgaki, Appl. Phys. Express 2, 036502 (2009).

10. R. A. Marsh, R. A., F. Albert, S. G. Anderson, G. Beer, T. S. Chu, R. R. Cross, G. A. Deis, C. A. Ebbers, D. J. Gibson, T. L. Houck, F. V. Hartemann, C. P. J. Barty, A. Candel, E. N. Jongewaard, Z. Li, C. Limborg-Deprey, A. E. Vlieks, F. Wang, J. W. Wang, F. Zhou, C. Adolphsen and T. O. Raubenheimer. Physical Review Special Topics: Accelerators and Beams 15(10): 102001 (2012)

11. M. A. Prantil, E. Cormier, J. W. Dawson, D. J. Gibson, M. J. Messerly and C. P. J. Barty. Optics Letters 38 (17): 3216-3218 (2013)

12. M. Y. Shverdin, M. Y., I. Jovanovic, V. A. Semenov, S. M. Betts, C. Brown, D. J. Gibson, R. M. Shuttlesworth, F. V. Hartemann, C. W. Siders and C. P. J. Barty. Optics Letters 35(13): 2224–2226. (2010)

Compact Gamma-Beam Source
for Nuclear Security Technologies

P. Gladkikh*[1] and J. Urakawa[2]

[1] *NSC KIPT, 1, Academicheskaya St.*
Kharkov, 61108, Ukraine
[2] *KEK, Accelerator, 1-1, Oho*
Tsukuba, Ibaraki 305-0801, Japan
* *gladkikh@kipt.kharkov.ua*

A compact gamma-beam source dedicated to the development of the nuclear security technologies by use of the nuclear resonance fluorescence is described. Besides, such source is a very promising tool for novel technologies of the express cargoes inspection to prevent nuclear terrorism. Gamma-beam with the quanta energies from 0.3 MeV to 7.2 MeV is generated in the Compton scattering of the "green" laser photons on the electron beam with energies from 90 MeV to 430 MeV. The characteristic property of the proposed gamma-beam source is a narrow spectrum (less than 1%) at high average gamma-yield (of $10^{13} \gamma/s$) due to special operation mode.

Keywords: Nuclear waste managements; nuclear resonance fluorescence; nuclear security; Compton storage rings.

1. Introduction

The nuclear resonance fluorescence (NRF) is one of the most promising methods of the novel technologies of the nuclear security and nonproliferation. This method consists in the excitation of transition between nuclear states; the detectors then register re-emitted gamma-rays. Spectrum of the nuclear transitions is unique for each isotope, i.e. is the isotope "fingerprint". For this method, a tunable source of the gamma–rays ranging from approximately 0.5 MeV to several MeV, with high spectral intensity of gamma–ray beam – flux about 10^{10} gamma's per second per 0.1 % of spectral width – is necessary. Besides, a narrow gamma-beam spectrum is also desirable, because such a source feature determines the signal to noise ratio in the measurements.

Modern level of both accelerator and laser technologies allows developing of such a source based on Compton scattering. Compton backscattered radiation possesses some features important for application of this radiation for the NRF. These are: high spectral brightness, much above most of conventional available sources, which reduces time of monitoring down to economically attractive one; a specific spectrum with the maximum close to high energy steep cutoff – dilution of the width of subtracted spectra can be as narrow as doubled energy spread in the electron bunches scattering off the laser pulses.

At present three types of the Compton based sources are under developing. In sources, based or linacs[1,2] or energy recovery linacs (ERL[3]) an electron beam is only once used for the gamma-beam generation. The electron beam emittance of modern linacs is very low, as well as the beam energy spread. These features lead to a narrow gamma-beam spectrum. The source disadvantage is a low average beam current, that leads to a quite low gamma-beam intensity.

Much more gamma-beam intensity can be generated in the Compton storage rings (CSR[4,5]), where electron beam generates the scattered gamma-quanta at every beam turn. Serious drawback of such type sources is the electron beam degradation into storage rings. Due to quantum fluctuations of the radiation and intra-beam scattering both emittance and energy spread of the electron beam increase comparably to the beam of the injector. In turn, it can lead to the considerable spectrum widening.

The main design idea of the proposed source is the combination of the advantages of both ERL–and–CSR–based schemes. It may be used in the source, consisting of the pulse "warm" linac and CSR by the proper choice of their parameters such as beam intensity, repetition rate, generation time etc.

2. CSR lattice

Taking into account the aforesaid, we have designed the lattice of the compact Compton ring, which meet requirements of the source for the nuclear resonance fluorescence. Its layout is presented in Fig 1. To solve the described problems we propose the gamma-source composed of the high-repetition (up to 300 Hz), low emittance linac-injector with large bunch population and low beam energy spread and compact storage ring with quite low natural beam emittance. To avoid significant widening of the gamma-beam spectrum during gamma-generation we assume pulse operation mode, when generation time is much less than pulse repetition cycle of the linac. From the point of the ring compactness we suppose to use the high bending field (of 18 T). Besides, we consider the "green laser" for the gamma-beam generation that enables decreasing of the electron beam energy. Gamma-yield is proportional to the collision frequency, that is why we suggest multi-bunch operation mode of both storage ring and optical cavity with bunch-to-bunch spacing equaling to the doubled rf-wavelength.

Both high bending field and long optical cavity enable to design the ring with head-on collision. The main physical advantage of this ring is weak dependence of the gamma-yield on the electron bunch length (in other words, on the scattering intensity). From the technology point the ring advantages are also marked: in this case the ring operation with two collision points is possible, as it is presented in Fig 1 (roughly speaking, it allows us to double the ring capability). The main ring parameters are listed in Table 1.

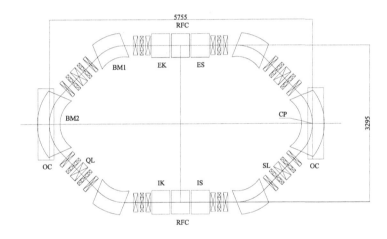

Fig. 1. CSR layout. BM1, BM2: bending magnets; QL, SL: quadrupoles and sextupoles; RFC: rf-cavity; OC: optical cavity; IS, IK: injection septum and kicker; ES, EK: extraction septum and kicker.

3. Linac-injector

Proposed gamma-beam source requests a high quality multi-bunch electron beam injection without the injection errors; an energy spread of an electron beam must be less or order 0.1 % to obtain a narrow gamma-beam spectrum[6]. As a consequence, multi-bunch and single-bunch beam loading compensation are necessary in the linac. Also, precise orbit control is important to avoid the injection errors. Fig. 2 shows the linac layout.

Table 1.   Main source parameters.

Parameter	Value
Electron beam energy range, MeV	90–450
Ring circumference, m	14.760
Linac length, m	25
Gamma-rays energy range, MeV	0.3–7.2
Gamma-yield, $photons/s/CP$	$\sim 1 \times 10^{13}$
Spectral density of collimated spectrum, $photons/keV/s$	$0.5 \times 10^9 - 2 \times 10^{10}$
Collimated spectrum FWHM, $\%$	~ 0.7
Repetition rate, Hz	300
Electron bunch charge, nC	0.5
Bunch–to–bunch spacing, ns	3.077
Number of electron bunches	16
Laser flash energy, $mJ/bunch$	10–25
Laser photons energy, eV	2.33
Laser waist, μm	30
Number of stacked laser bunches into optical cavity	6

Fig. 2.   Linac layout.

The designed linac consists of:

- 2.6 cell RF–gun;
- six 3 m long traveling wave accelerating structures;
- chicane just after the RF–gun and 45–degrees bending magnet just after the first 3 m structure for the beam diagnostics;
- 7 klystrons and modulators;
- 3 sets of phase to amplitude controllers for multi-bunch beam loading compensation;

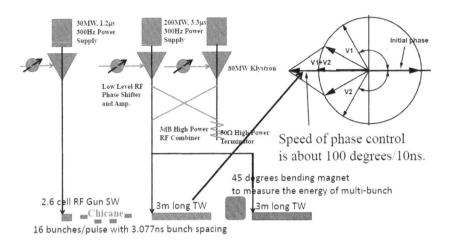

Fig. 3. Fast and precise phase control for the multi-bunch beam loading compensation.

- magnet system for the beam focusing and steering;
- the beam diagnostics and control system.

The scheme[7] for the phase to amplitude beam loading compensation for the upstream region is presented in Fig. 3. To get the high accelerating gradient of 36 MV/m we have to set-up SLED (SLAC Energy Doubler) which is pulse compressor.

The beam time structure and complete RF unit are shown in Fig 4.

4. Simulation of the Compton scattering

Many simulations of the Compton scattering in the chosen CSR–lattice have been carried out to obtain the most appropriate parameters of the gamma-beam generation. The final results at the minimal and maximal electron beam energy are presented in Figs. 5–7. As it is seen in these figures, the width of the collimated beam spectrum under small opening angle is determined by the energy spread of the electron beam.

5. Conclusions

We propose compact gamma-beam source, which joint the advantages of both CSR and ERL schemes together. The spectrum width is practically

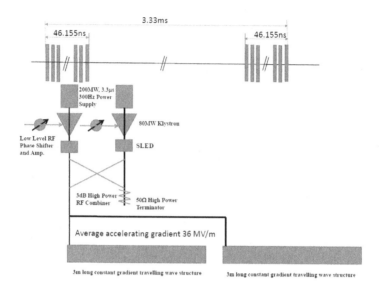

Fig. 4.   Time structure of multi-bunch beam and RF-unit of linac.

determined by the energy spread of the injected electron beam. Use of the head-on collision allows us to improve the source capabilities considerably. Presented source can be effectively used for the cargo inspection, nuclear waste management, measure of the melted fuel in core at the nuclear accidents etc.

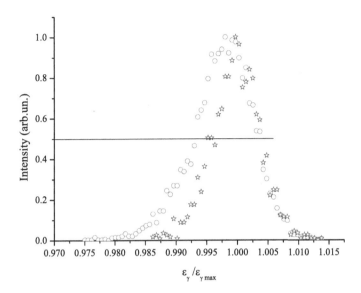

Fig. 5. Collimated spectrum at the gamma-beam energy $\varepsilon_\gamma^{\max} = 0.5\,\text{MeV}$ (electron beam energy $E_e = 100\,\text{MeV}$). The data points are for $0.4\,\text{mrad}$ ($\circ$) and $0.2\,\text{mrad}$ ($\star$).

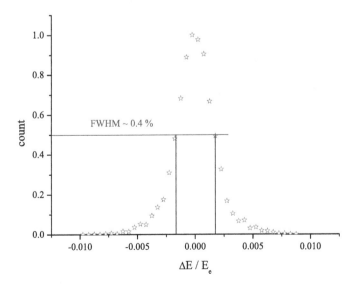

Fig. 6. Electron beam energy spread (electron beam energy $E_e = 100\,\text{MeV}$).

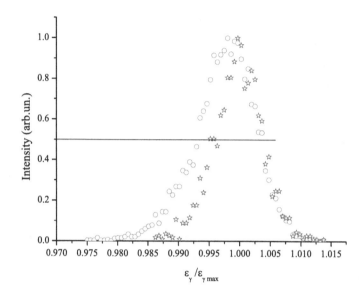

Fig. 7. Collimated spectrum at the gamma-beam energy $\varepsilon_\gamma^{\max} = 7.2\,\mathrm{MeV}$ (electron beam energy $E_e = 450\,\mathrm{MeV}$). The data points are for $0.1\,\mathrm{mrad}$ ($\circ$) and $0.05\,\mathrm{mrad}$ ($\star$).

References

1. J. Pruet, D.P. McNabb, C.A. Hagmann, F.V. Hartemann and C.P.J. Barty, Detecting clandestine material with nuclear resonance fluorescence, *J. Appl. Phys.*, **99**, 123102-1-11 (2006).

2. J. Urakawa, Development of a compact X-ray source based on Compton scattering using a 1.3 GHz superconducting RF accelereating linac and a new storage cavity, *Nucl. Instrum. Methods*, **A 637**, S47–S50 (2011).

3. R. Hajima, T. Hayakawa, N. Kikuzawa, E. Minehara, Proposal of non-destructive radionuclide assay using a high-flux gamma-ray source and nuclear resonance fluorescence, *J. Nucl. Sci. Tech.*, **45**, 441–451 (2008).

4. E. Bulyak, P. Gladkikh, T. Omori, V. Skomorokhov, J. Urakawa, Compton ring for nuclear waste management, *Nucl. Instrum. Methods*, **A 621**, 105–110 (2010).

5. P. Gladkikh, Lattice and Beam Parameters of Compact Intense X-rays Source Based on Compton Scattering, *Phys. Rev. ST Accel. Beams*, **8**, 050702 (2005).

6. M. Fukuda, S. Araki, A. Aryshev, Y. Honda, N. Terunuma, J. Urakawa,

K. Sakaue, M. Washio, in *Proceedings of PAC2013*, **TUPMA01**, (Pasadena, CA USA, 2013).

7. S. Kashiwagi, H. Hayano, K. Kubo, T. Korhonen, S. Nakamura, T. Naito, K. Oide, S. Takeda, N. Terunuma, J. Urakawa, *Japanese Journal of Applied Physics*, **8A**, 5617–5622 (2004).

Developments of Optical Resonators and Optical Recirculators for Compton X/γ Ray Machines

A. Martens on behalf of the MightyLaser, ThomX collaborations and the

EuroGammaS association

LAL, Univ Paris-Sud, CNRS/IN2P3, Orsay, France
martens@lal.in2p3.fr
www.lal.in2p3.fr

Optical resonators and optical recirculators are key elements of Compton X/γ ray machines. With regard to their use in laser physics or in time-frequency metrology, these devices have to obey severe constraints when implemented in the vaccum of an electron accelerator. Our group has developed both types of devices. In this proceedings an original recirculator design, that was developed within the European proposal to the ELI-NP γ ray source call for tender, is described. This is an aberration free device which allows reciculating 32 times a short and high intensity laser pulse. It also allows synchronizing each of the 32 passes with the electron RF cavities within 100 fs. The second topic of these proceedings is a description of our R&D on optical resonators dedicated to laser-electron interactions. We have locked two different picosecond laser oscillators to the highest cavity finesse F=30000 ever reached in pulsed regime. We also designed and build a new kind of non-planar cavity, tetrahedron shape, providing circularly polarized eigen modes. This cavity was installed in the ATF accelerator of KEK and successfully used to produce a high gamma ray flux. Thanks to an original fibre amplifier, we succeed in stacking 100 kW of average power inside the cavity.

Keywords: Compton sources; X/γ rays; optical resonators; optical circulators.

1. Introduction

Applications of Compton backscattering off electrons cover a wide range of photon energies from X-rays to high energy γ-rays[1]. The energy $E_\gamma \approx 4\gamma^2 \frac{hc}{\lambda}$ of the backscattered photons can be tuned either by changing the incident light wavelength λ or, more easily, by adapting the electron beam energy $E_e = \gamma m_e c^2$, where m_e is the electron mass, c the speed of light in vacuum and h the Planck constant. Three typical ranges of applications can be distinguished. Lower energies (X-rays) applications typically from a keV to less than 300 keV cover the medical (diagnosis and treatment)[2] and art history domains. Intermediate energies from 200 keV to 20 MeV cover the nuclear physics, from fundamental science experiments to nuclear waste management[3] and nuclear non-proliferation. High energy physics applications exist for photons above 20 MeV, among them electron beam polarimetry, photon-photon collider and production of polarised positrons.

These radiation sources require careful designs of optical systems and generally appear challenging. The Laboratoire de l'Accélérateur Linéaire (LAL, Orsay, France) is currently engaged in three experiments and projects. The MightyLaser experiment[4], installed at KEK (Tsukuba, Japan), is a R&D optical cavity designed for circularly polarized photon production[5,6] that are necessary to produce longitudinally polarised positrons[7], in collaboration with the CELIA (Bordeaux, France), KEK (Tsukuba, Japan), Hiroshima University (Japan) LMA (Lyon, France) laboratories and the Posipol Collaboration. This is part of the R&D in view of a potential future Linear Collider (LC)[8,9]. The setup, preliminary results and first lessons learned from this experiment will be discussed in these proceedings. This R&D effort is pursued towards the construction of the optical cavity for the ThomX project[10], an X-ray source being built at LAL. A complementary strategy to optical cavities, involving optical circulators, is developed for the gamma source of the Extreme Light Infrastructure - Nuclear Physics (ELI-NP) project[11] within the EuroGammaS collaboration[12]. Both of these ongoing projects will also be briefly discussed in these proceedings.

2. Challenges and strategies

The requirements on the photon source for the above-mentioned ongoing projects are different. Apart from the different energy range, corresponding to different electron beam energies, the requested flux for ThomX is of 10^{11} to 10^{13} photons per second (ph./s) while the main constraint for the ELI-NP project is the time-averaged spectral density, that is required to be above $5000 \text{ s}^{-1}\text{eV}^{-1}$ at 10 MeV. Since a high flux is required for the ThomX project, a high repetition rate is needed and thus an electron ring is being constructed[13]. In order to reach a high enough flux of photons, an optical enhancement cavity is designed to magnify the otherwise limited average-power of the laser. On the other hand, the choice of a linear electron accelerator is driven by the high spectral purity of the photon flux required for the ELI-NP project. In this case the collision rate is too low to envisage the use of a resonant cavity. An optical circulator is thus being designed to maximise the flux of the source while preserving its spectral properties[14].

These two different strategies, cavity and circulator, imply the precise control of several effects limiting the performance of the apparatus. The importance of these limitations depends on the strategy employed. Among them the optical oscillator phase noise, the thermal effects and the limitations arising from imperfect alignment and synchronisation are of primary importance. These will be discussed in these proceedings.

3. The MightyLaser experiment

3.1. *Experimental setup*

The MightyLaser optical cavity[4] is installed at KEK (Tsukuba, Japan) and is designed for the R&D towards polarized positron sources based on Compton backscattering for a LC. It is thus necessary to produce circularly polarized photons that will impinge a thin conversion target to produce positrons that are captured and accelerated. The circularly polarised γ source is produced thanks to a non planar (tetrahedron) geometry of the cavity[6]. This four-mirror cavity which allow an independent optimisation of the optical round trip and the waist size, is composed of two plane and two spherical mirrors introducing also general astigmatism[15] in the cavity.

The optical oscillator used to feed the optical cavity is a 100 fs Yb Orange oscillator from MENLO[16], with a repetition frequency of approximately 178 MHz and an output average-power of 20 mW. It is amplified to 50 W by means of the Chirped Pulse Amplification (CPA) technique[17]. A strechter unit is followed by an electro optic modulator (EOM) necessary for the Pound-Drever-Hall (PDH) technique[18] used to lock the laser frequency comb on the modes of the cavity. The amplification is performed in three stages and the pulse is then compressed down to approximately 10 ps for the collisions with the electron beam. A numerical feedback with a bandwidth of few MHz is implemented to keep the laser locked on the cavity and the cavity on the electron accelerator RF.

3.2. *Results*

A first set of results were obtained in March 2011[4,19] with a cavity of finesse 3000 with approximately 10 W of incident laser power. Approximately 0.2 kW were stored in the cavity in continuous regime giving a flux of approximately 10^6 ph./s with a OneFive Origami laser[20].

New mirrors were installed in 2013 to improve the finesse of the cavity by an order of magnitude. With approximately 50 W of incident laser power, slightly more than 100 kW were obtained in transient regime. This number reduces to 40 kW in continuous regime due to thermal effects observed both in the optics outside and inside the cavity. A few hundreds of Compton backscattered photons were obtained per beams crossing.

3.3. *Discussion*

During this experiment several effects that limit the power stored in the cavity were identified. First thermal effects outside the cavity, in the Chirped Volume Bragg Grating (CVBG) compressor, were observed with 50 W of incident power. A laminar flow was thus installed to slightly mitigate the effect without large improvement in performance.

Another limitation arises from thermal effects in the cavity due to absorption of power in the substrate of the mirrors. Indeed the temperature gradient induces a change in the sagitta of the mirrors due to non vanishing thermal expansion and finite thermal conductivity. This modification induces a change in the transverse size of the beam on the transmission mirror, thus reducing again the coupling of the input laser with the mode of the cavity. This additional loss of coupling appears progressively after the lock of the laser on the cavity with a typical time constant of several hundreds of ms and was clearly observed on the intensity transmitted behind a plane mirror of the cavity. This effect can be strongly suppressed[21] by using Ultra Low Expansion (ULE) substrates instead of Fused Silica, as was used for the MightyLaser cavity.

Finally, effects related to the phase noise of the amplified laser are also reducing the coupling of the incident laser to the cavity. These three effects must be further worked on to improve the accumulated power in the cavity for future projects.

4. The ThomX project

ThomX[10] is a machine that is being build at Laboratoire de L'accélérateur Linéaire, Orsay, France and will be commissioned in a couple of years. This X-ray machine is designed for art history and medical applications and build in collaboration with experts of these fields. The electrons are stored in a ring and the beam energy will be tunable between 50 MeV and 90 MeV to change the X-rays energy. The charge of the bunch is expected to be of 1 nC with normalized transverse emittance below 1 mm.mrad and energy spread of few per mile.

The collision rate between the electron beam and the laser beam will be of 17.8 MHz while the cavity Free Spectral Range will be of 35.6 MHz to limit its size and mitigate potential stability issues with an effectively long cavity. The optical, Fabry-Pérot type, cavity has a planar geometry for the production of a linearly polarized photon beam.

X-ray experiments, located at 10 meters away from the Compton interaction point, will benefit from a high flux of photons, a rather large beam divergence of 10 mrad in absence of a diaphragm. The spectral bandwidth is expected to be in the 1 to 10 % range when a diaphragm is employed.

Some of the advantages of such a machine compared to synchrotron radiation sources are its compactness and relatively limited cost.

5. The ELI-NP gamma beam source

As part of the ELI-NP project[22], two γ-ray sources with distinct energy ranges are foreseen. The constraints on the properties (mainly flux and spectral density) of these sources are stringent. In particular the required spectral density strongly limits the possible choices of accelerator technologies.

X-band and ERL are interesting technologies to satisfy the needs of ELI-NP but were judged too risky by the EuroGammaS association given the time constraints imposed for the construction of the machine. Thus the EuroGammaS association will build an hybrid C and S band electron Linear Accelerator (LINAC)[23]. The two interaction points will have a steplessly tunable energy over two orders of magnitudes. Thus two interaction points will be build. Photons produced at the first Compton interaction point (IP1) will have an energy tunable between 0.2 and 3 MeV. Photon energies of at least 20 MeV must be reached at the second interaction point (IP2). In order to achieve the large spectral density of 5000 ph./(s eV), that is approximately two orders of magnitude larger than the state of the art, a collimator is required to absorb Compton backscattered photons emitted off electron-axis. This will ensure to preserve the monochromaticity of the source below 0.5% of the average beam energy. In order to reach a large enough flux it is necessary to have several bunches in the electron beam. The LINAC will thus deliver electron beams to the two IPs with tunable energies of approximately 250 MeV and 600 MeV. Trains of 32 bunches, of 250 pC each and separated by 15.6 ns, will collide every 10 ms at the IPs.

This scheme for the LINAC essentially imposes the use of an optical circulator rather than a resonator, which would be rather inefficient given the ratio of the average collision rate and reasonably achievable resonator free spectral ranges (FSR) of few tens of MHz. In this scheme a large interaction laser energy, typically of 200 mJ at 515 nm is required. This will be achieved thanks to the use of an Yb oscillator amplified with the Chirped Pulse Amplification (CPA) technique.

The ideas of a confocal circulator[24] or an Herriot cell[25] cannot be used since they either have variable crossing angle, which spoils the energy resolution, or a wild focusing region, which reduces the instantaneous luminosity and the quality of the source spot averaged over the 32 bunch crossings. Thus the solution envisaged is made of two parabolic mirrors in between of which are placed 31 optical-invariant mirror-pair systems (MPS) on an helicoid. These MPS allow to swap from one interaction plane (containing the electron beam and the laser beam) to the next one. The 32 interaction planes are rotated by an azimuthal angle of approximately 8° around the electron beam axis, after an optimisation of the parameters of the whole optical system thanks to full simulations[14].

The stringent constraints on the 0.2 mrad divergence of the photon beams and the size of the beam spots of few tens of microns at 10 m from the IPs require to have an excellent alignment of the optical circulator. Effects of misalignments have been studied and show that with 20 microns of residual misalignment, the average relative spectral-density, with respect to a perfect optical system, is of approximately 0.7. Such a loss in performance would allow to reach the requirements of the ELI-NP project. But given that a new kind of optical system is being build and that other losses in terms of performances can arise from several places, it is believed to be dangerous not to further optimize the alignment. Such an optimisation, based on realistic simulations, was performed thanks to a dedicated algorithm by adjusting five degrees of freedom (dof) on one of the parabola and the two additional tilts of the injection mirror. The result of this optimisation demonstrates that 0.98 relative spectral-density is reached in average with a small dispersion of 0.01 corresponding to different initial misalignment conditions. The performance of the system is thus optimized with this algorithm. It has also been checked that the procedure is robust against larger initial misalignments of 100 μm. Experimental inputs to the alignment procedure will be provided by a dedicated low energy laser that will be forced to circulate in the optical system during dedicated calibration runs. A thin pellicle coupled to an intensified Coupled Charge Device (CCD) via an imaging system will allow the measurement of the position of the 32 spots at the interaction point.

Given the few picoseconds length of both electron and laser beams, a synchronisation of the 32 laser passes and the 32 electron bunches is typically required to be better than 200 fs. Each of the laser passing at the interaction point will be synchronised to the electron beam by fine adjustment of each of the MPS angles, thus varying with a good sensitivity the path length of the laser in the optical circulator. A similar setup will be used to allow the synchronisation on the accelerator RF system by using the above-mentioned alignment laser that will be split in two beams. One of these will circulate in the optical system as will do the interaction laser in normal operation after picking the first pulse to avoid the superimposition of several pulses on the camera, while the other one will follow the trajectory of the electrons in the circulator. Finally it is believed that with such a system the requirements of the ELI-NP project will be reached.

6. Conclusion

The LAL group is actively collaborating in several experiments and projects dedicated for the production of Compton X and γ-ray sources. Two distinct and complementary strategies are pursued depending on the requirements for theses radiation sources. The first solution, designed to maximise the flux, is based on resonant cavities. We obtained up to 100 kW in transient regime and 40 kW for several hours of operation in an experiment producing backscattered Compton gamma rays. The second solution, designed to maximise the spectral density, is based on optical circulators. Improvements of few orders of magnitude compared to the state of the art are expected with the gamma source of the ELI-NP project.

Acknowledgments

The ThomX project is supported by the program "Investing in the future" ANR-10-EQOX-51. Work also supported by grants from Région Ile-de-France.

References

1. P. Sprangle *et al.*, Tunable, short pulse hard X-rays from a compact laser synchrotron source, *Journal of Applied Physics* **72**, 5032 (1992).
2. F. E. Carroll, Tunable monochromatic X rays: a new paradigm in medicine, *Am. J. Roentgenol.* **179**, 583 (2002).
3. R. Hajima *et al.*, Proposal of nondestructive radionuclide assay using a high-flux gamma-ray source and nuclear resonance fluorescence, *J. Nucl. Sci. Technol.* **45**, p. 441451 (2008).

4. J. Bonis *et al.*, Non-planar four-mirror optical cavity for high intensity gamma ray flux production by pulsed laser beam Compton scattering off GeV-electrons, *JINST* **7**, p. P01017 (2012).

5. H. Kogelnik and T. Li, Laser beams and resonators, *Appl. Opt.* **5**, 1550 (Oct 1966).

6. F. Zomer *et al.*, Polarization induced instabilities in external four-mirror fabry-perot cavities, *Appl. Opt.* **48**, 6651 (Dec 2009).

7. S. Araki *et al.*, Design of a polarised positron source based on laser Compton scattering, arXiv:physics/0509016, (sep, 2005).

8. C. Adolphsen *et al.*, The International Linear Collider Technical Design Report - Volume 3.I: Accelerator R&D in the Technical Design Phase, arXiv:1306.6353 (jun, 2013).

9. M. Aicheler *et al.*, A Multi-TeV Linear Collider Based on CLIC Technology: CLIC Conceptual Design Report, `http://dx.doi.org/10.5170/CERN-2012-007` (oct, 2012).

10. A. Variola *et al.*, ThomX Conceptual Design Report, hal.in2p3.fr/in2p3-00448278 (feb., 2010).

11. S. Gales, Nuclear Science and applicataions with next generation of high power lasers and brilliant low energy gamma beams at ELI-NP, in these proceedings, (2014).

12. `http://www.e-gammas.com/`.

13. Z. Huang and R. D. Ruth, Laser-electron storage ring, *Phys. Rev. Lett.* **80**, p. 976 (1998).

14. K. Dupraz *et al.*, Design and optimization of a highly efficient optical multipass system for γ-ray beam production from electron laser beam Compton scattering, *Phys. Rev. ST Accel. Beams* **17**, p. 033501 (Mar 2014).

15. J. A. Arnaud and H. Kogelnik, Gaussian light beams with general astigmatism, *"Appl. Opt."* **8**, p. 1687 (1969).

16. `http://www.menlosystems.com/`.

17. D. Strickland and G. Mourou, Compression of amplified chirped optical pulses, *Opt. Commun.* **56**, 219 (1985).

18. R. Drever *et al.*, Laser phase and frequency stabilization using an optical resonator, *Appl. Phys. B* **31**, 97 (1983).

19. T. Akagi *et al.*, Production of gamma rays by pulsed laser beam Compton scattering off GeV-electrons using a non-planar four-mirror optical cavity, *JINST* **7**, p. P01021 (2012).

20. `http://www.onefive.com/origami.html`.

21. H. Carstens *et al.*, Cavity-enhanced 196 kW average-power infrared

pulses, in *Advanced Solid-State Lasers Congress*, *Advanced Solid-State Lasers Congress* (Optical Society of America, 2013).

22. http://www.eli-np.ro/.

23. L. Serafini, Challenges of a new proposed Compton Source for the ELI-NP Europeans Infrastructure, in these proceedings, (2014).

24. A. Rollason *et al.*, Multipass optical cavity for inverse compton interactions, *Nucl. Instrum. Methods Phys. Res. A* **526**, 560 (2004).

25. D. Herriott *et al.*, Off-axis paths in spherical mirror interferometers, *Appl. Opt.* **3**, 523 (Apr 1964).

Low-Lying "Pygmy" Dipole Resonances and Strength Functions

V. Werner

Institut für Kernphysik, TU Darmstadt, Schlossgartenstraße 9, D-64289 Darmstadt, Germany
Wright Nuclear Structure Laboratoy, Yale University, P.O. Box 208120, New Haven, CT 06520-8120, USA
vw@ikp.tu-darmstadt.de

N. Cooper

Wright Nuclear Structure Laboratoy, Yale University, P.O. Box 208120, New Haven, CT 06520-8120, USA

P. M. Goddard, R. S. Ilieva

Wright Nuclear Structure Laboratoy, Yale University, P.O. Box 208120, New Haven, CT 06520-8120, USA
Department of Physics, University of Surrey, Guildford, GU2 7XH, UK

P. Humby

Wright Nuclear Structure Laboratoy, Yale University, P.O. Box 208120, New Haven, CT 06520-8120, USA
University of Richmond, Richmond, VA 23173, USA
Department of Physics, University of Surrey, Guildford, GU2 7XH, UK

N. Pietralla

Institut für Kernphysik, TU Darmstadt, Schlossgartenstraße 9, D-64289 Darmstadt, Germany

Recent investigations into dipole resonances below the neutron separation threshold have focused on characterizing the properties of the so-called Pygmy Dipole Resonance. The amount of extra PDR strength on top of a GDR tail depends largely on the choice, or the method of extraction of photon strength functions. Whereas most experimental searches for the PDR were performed on spherical nuclei, the present work focuses on recent experiments on ^{76}Se and ^{76}Ge, on the virge of deformation.

Keywords: Pygmy dipole resonance; nuclear resonance fluorescence; photon strength functions.

1. Introduction

Electric dipole excitation strengths around 6-8 MeV are often referred to as the pygmy dipole resonance (PDR)[1,2], an electric dipole resonance on the tail of the giant dipole resonance (GDR). The PDR is often interpreted as a collective oscillation of a neutron skin against a proton-neutron core[3,4]. If this is the origin of additional E1 excitation strength on the low-energy GDR tail, the neutron excess should be correlated to the magnitude of this strength. The PDR may have significant impact in nuclear astrophysics regarding nucleogenesis through rapid neutron capture[5].

At lower energies, typically about 2-4 MeV, other types of dipole excitation modes are located. In near-spherical systems, these are two-phonon excitation. One is formed by coupling the first quadrupole (2_1^+) and octupole (3_1^-) excitations, hence, an electric dipole state, which is a member of the resulting two-phonon multipelt. The other stems from the coupling of the proton-neutron symmetric 2_1^+ state to its one-phonon mixed-symmetric $2_{1,\mathrm{ms}}^+$ counterpart, where not all protons and neutrons are in phase, forming a multiplet of states (see Ref.[6]). The $J^\pi = 1^+$ member of this multiplet evolves into the well-known scissors mode in deformed nuclei. Impact of scissors mode decays on model parameters with respect to neutrino-less double-beta decay has recently been demonstrated[7].

To date, the PDR has mostly been searched for in spherical nuclei on or near closed proton or neutron shells, and E1 excitation has been measured up to the neutron separation threshold in experiments using the nuclear resonance fluorescence (NRF) technique[8]. In particular, nuclei of interest have been excited from their ground states by the use of real photons, produced either by bremsstrahlung or Compton backscattering. The latter γ-ray beams have the advantage of being fully polarized, hence, allowing to distinguish between E1 and M1 excited states which may coexist in the same energy range, like the E1 and M1 two-phonon states at low energies, or the PDR and M1 spin flip resonance states below or at the neutron separation threshold[9,10]. In the following we will focus on the photon scattering techniques, and discuss recent results on a study of the double-beta decay partners ^{76}Se and ^{76}Ge, which have significantly different quadrupole deformations.

2. Experiments at the S-DALINAC

The recirculating, superconducting Darmstadt linear accelerator (S-DALINAC) is the central infrastructure of the institute for nuclear physics at TU Darmstadt. The injector of the S-DALINAC is used for the production of bremsstrahlung γ-ray beams up to 10 MeV. Also higher-energy beams, making use of the main accelerator, can be produced at a different location. The main setup for NRF, the Darmstadt high-intensity photon setup (DHIPS)[11] behind the injector starts with the bremstarget, for which typically a thick copper place is used. Depending on the required beam energy, considering the neutron separation threshold of the bremstarget, other materials can be used to optimize the γ-ray production. A continuous bremsstrahlung spectrum is produced, up to the energy of the incident, mono-energetic electron beam.

The photon beam is then collimated through a long copper collimator and impinges on the NRF target - typically isotopically enriched material on the order of few grams. The target is surrounded by three Compton-suppressed HPGe detectors, two at 90° and one at 130° with respect to the beam axis. The angles are chosen to allow a distinction between $J = 1$ and $J = 2$ states in even-even nuclei from the respective angular distributions of emitted γ-rays. A second NRF targets and additional detectors can be placed downstream. All detectors are placed inside a massive lead housing, in order to be shielded from radiation background from the accelerator hall.

Eigenstates in the target nuclei are resonantly excited by the incident γ-rays, and subsequently emit radiation from their decays back to the ground state or to lower-lying excited states. In most cases only decays to the lowest-lying states can be observed, since non-resonantly scattered photons result in an exponentially increasing background toward lower energies. Cross sections are determined relative to known calibration standards like ^{11}B or ^{27}Al, which are added to the target.

3. Experiments at HIGS

Whereas bremsstrahlung experiments allow for spin assignments and the extraction of absolute cross sections, the determination of parities usually requires the use of Compton polarimetry with its limited sensitivity, especially at high γ-energies. An alternative is the use of polarized beams, such as delivered at the high-intensity gamma-ray source (HIGS)[12] at the Triangle Universities Nuclear Laboratory (TUNL). Intense γ-ray beams are produced by the process of Compton backscattering of laser photons from relativistic electrons. The laser light is produced by the same electron bunches within a storage ring in a free electron laser system. The polarization of the laser light is maintained in the Compton backscattering process, hence, the resulting γ-ray beams are almost fully polarized (restricted only by geometry). Another advantage of HIGS γ-ray beams is the relatively narrow bandwidth of the system. With collimators used in our experiments, typically an energy spread of about 3 % was achieved.

With such a quasi-monoenergetic γ-ray beam, either individual states, or small groups of states can be excited and their decays to lower-lying states be detected[13]. Background toward low energies does exist, however, is much reduced compared to the background in a bremsstrahlung experiment. Therefore, the potential of directly observing decays to lower-lying excited states, in addition to decays to the ground state, is higher at HIGS. The detector setup used in our experiments consisted of two detectors in the plane spanned by the polarization vector and the beam axis (horizontal), and two detectors in the perpendicular plane (vertical), all at 90° relative to the beam axis. This polarimeter setup allows to directly measure parities of dipole excited states, which will emit γ-rays from the decay to the ground state into the horizontal plane in case of positive parity (M1 excitations), or into the vertical plane in case of negative parity (E1 excitations). The corresponding angular distributions are shown in Fig. 1. Defining an asymmetry as

$$P = Q \cdot \Sigma = Q \cdot \frac{N_{hor} - N_{ver}}{N_{tot}} , \qquad (1)$$

M1 excited states have ideally $P = 1$ and E1 excited states $P = -1$. These values are only attenuated by the factor Q due to solid angles of the detectors, and N_x are the intensities observed in the horizontal, vertical, or all detectors, respectively.

In addition, cross sections of newly observed states can be deduced relative to those of neighboring states which are covered in the same beam setting, and the cross sections of which are already known, e.g., from bremsstrahlung experiments. To achieve this, the photon flux at the target position is simulated using GEANT4, compared to a spectrum taken with an additional detector at zero degrees, which is directly by an (attenuated) beam, and then scaled to the known cross sections.

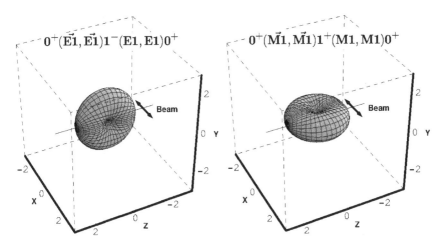

Fig. 1. Angular distributions of γ-rays from the decay of M1 and E1 excited states, respectively, in even-even nuclei to the ground state after excitation from the ground state with polarized photons.

4. Results on $A = 76$ Se and Ge

The isotope ^{76}Se has been measured at the S-DALINAC and at HIGS in the energy range of approximately 4 - 9 MeV, using the methods sketched above. From the bremsstrahlung experiment, absolute cross sections were obtained. However, since electric and magnetic dipole excitations coexist in this energy region, parity information had to be obtained at HIGS. Figure 2 shows the measured asymmetries, proving that almost all observed states have negative parity. At HIGS, more states have been observed than at the S-DALINAC, and combining both data set results in a rather complete E1 strength distribution up to 9 MeV - limited at higher energies only by the detection limit in this experiment. The complete data from these experiments has recently been published in Ref.[14,15].

In both experiments, only few transitions to lower-lying excited state have been observed. However, peaks corresponding to transitions from low-lying 2^+ states occured in the spectra. These states have been populated through non-observed decay cascades from excited states around the beam energy. Therefore, cross sections that were obtained from ground state decays only need to be corrected by those alternate decay paths. Since these branching decays can stem from any state within the beam profile, this correction can only be done in average for each beam setting. This results in the cross section distribution shown in Fig. 3, which is adopted

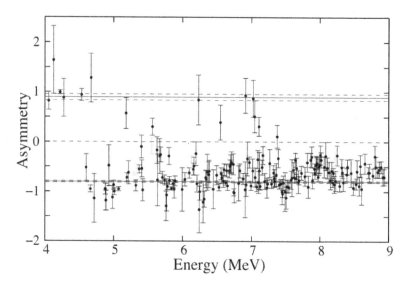

Fig. 2. Measured asymmetries of states in ^{76}Se as a function of excitation energy. Green points are potential doublets.

from Ref.[15], including also data from (γ,n) experiments above the neutron separation threshold[16].

The ongoing analysis of ^{76}Ge data, which was measured in analogous experiments, yields a similar E1 cross section distribution, however, the structure at about 7 MeV appears to be missing. Overall, the connection to (γ,n) data, covering the GDR, is rather smooth in both nuclei. An obvious problem for quantifying the amount of E1 strength on top of the tail of the GDR is the parametrization of the GDR itself, that is the application of a good photon strength function.

At present, statistical calculations are performed in order to investigate this point. Various strength functions are used as input into these calculations, involving Monte Carlo simulations of level distributions and decay widths, with the input of literature parametrizations of level densities, an Ansatz for typically Lorentzian type strength functions, and statistical (Porter Thomas) fluctuations. The validity of the Axel-Brink hypothesis[17,18] is assumed, which states that resonances are built on excited states the same way as on the ground state, which impacts the decay branches of excited states. In a first series of such calculations we find little enhancement of E1 strength above the low-energy tail of the GDR for ^{76}Se.

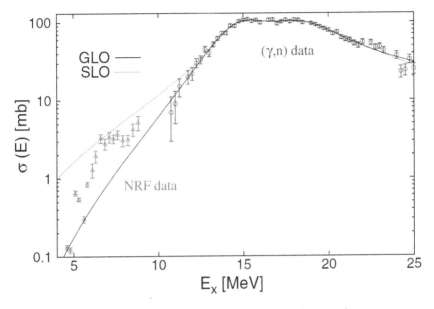

Fig. 3. Photo-excitation cross sections from the present work up to the neutron separation energy, and (γ,n) data combined. Data are compared to a Standard Lorentzian and a Generalized Lorentzian fit as indicated.

Therefore, a question is whether the structure around 7 MeV is due to a neutron skin resonance, or can be attributed to statistical fluctuations and detection limits, which would result in less observed strength at high energies. This effect has been included in the statistical calculations, giving reasonable agreement with data. As mentioned above, for ^{76}Ge, at the present status of analysis, we observe no such structure.

5. Conclusions

We tested the dipole response of neutrino-less β-decay candidate pair ^{76}Se and ^{76}Ge using a powerful combination of NRF experiments with bremsstrahlung and Compton-backscattering beams. Continuous bremsstrahlung beams allowed for a coarse scan of dipole excited states and measurement of spins and absolute cross sections, whereas the quasi-monoenergetic beams gave access to parity information and information about the decay behavior of E1 excited states. A structure in the cross section distribution of ^{76}Se has been found around 7 MeV, the orgin of which is still under investigation. Preliminary analysis of ^{76}Ge does not show any such structure. The analysis will be completed, including low-lying two-phonon states and their decay behavior, to be compared to collective and microscopic models.

Acknowledgments

The authors acknowledge the help from the experimental groups from TUNL, EMMI/GSI, Yale University, University of Kentucky, TU Darmstadt, and NIPNE in performing the experiments. We are grateful for fruitful discussions with G. Rusev. Support from DFG under Grant No. SFB634, and U.S. DOE under Grant No. DE-FG02-91ER-40609, and the UK STFC under Grant Nos. ST/J500768/1, ST/J00051/1, and ST/005528/1.

References

1. R.-D. Herzberg, P. von Brentano, J. Eberth, J. Enders, R. Fischer *et al.*, Phys. Lett. **B 390**, 49 (1997).
2. D. Savran, T. Aumann, and A. Zilges, Prog. Part. Nucl. Phys. **70**, 210 (2013).
3. P. Adrich, A. Klimkiewicz, M. Fallot, K. Borezky, T. Aumann *et al.*, Phys. Rev. Lett. **95**, 132501 (2005).
4. N. Tsoneva and H. Lenske, Phys.Rev. C **77**, 024321 (2008).
5. S. Goriely, Phys. Lett. **B 436**, 10 (1998).
6. N. Pietralla, P. von Brentano, and A. F. Lisetskyi, Prog. Part. Nucl. Phys. **60**, 225 (2008).
7. J. Beller, N. Pietralla, J. Barea, M. Elvers, J. Endres *et al.*, Phys. Rev. Lett. **111**, 172501 (2013).
8. U. Kneissl, H.-H. Pitz, and A. Zilges, Prog. Part. Nucl. Phys. **37**, 349 (1996).
9. C. Iwamoto, H. Utsunomiya, A. Tamii, H. Akimune, H. Nakada *et al.*, Phys. Rev. Lett. **108**, 262501 (2012).
10. G. Rusev, N. Tsoneva, F. Dönau, S. Frauendorf, R. Schwengner *et al.*, Phys. Rev. Lett. **110**, 022503 (2013).
11. K. Sonnabend, D. Savran, J. Beller, M. Büssing, A. Constantinescu *et al.*, Nul. Instrum. Methods A **640**, 6 (2011).
12. H. R. Weller, M. W. Ahmed, H. Gao, W. Tornow, Y. K. Wu *et al.*, Prog. Part. Nucl. Phys. **62**, 257 (2009).
13. N. Pietralla, Z. Berant, V. N. Litvinenko, S. Hartmann, F. F. Hikhailov *et al.*, Phys. Rev. Lett. **88**, 012502 (2002).

14. N. Cooper, F. Reichel, V. Werner, L. Bettermann, B. Alikhani *et al.*, Phys. Rev. C **86**, 034313 (2012).

15. P. M. Goddard, N. Cooper, V. Werner, G. Rusev, P. D. Stevenson *et al.*, Phys. Rev. C **88**, 06408 (2013).

16. P. Carlos, H. Beil, R. Bergre, J. Fagot, A. Leprêtre *et al.* Nucl. Phys. A **258**, 365 (1976).

17. D. M. Brink, Ph.D. thesis, Oxford University (1955).

18. P. Axel, Phys. Rev. **126**, 671 (1962).

NRF-Based NDA of Nuclear Material Using Monochromatic γ-Ray Beam

T. Shizuma*, T. Hayakawa, C.T. Angell, R. Hajima, F. Minato, K. Suyama and M. Seya

Japan Atomic Energy Agency,
Tokai, Ibaraki 319-1195, Japan
*shizuma.toshiyuki@jaea.go.jp

M. S. Johnson and D. P. McNabb

Lawrence Livermore National Laboratory,
Livermore, CA 94550, USA

Nuclear resonance fluorescence (NRF) is useful for nondestructive assay (NDA) of nuclear materials such as spent nuclear fuel. Counting precision of the NRF-based measurement system can be affected by background counts from self-activity of spent fuel and coherent scattering such as Rayleigh, nuclear Thomson, and Delbrück scattering. In this talk, the measurement principle and calculated uncertainties of the proposed detection system are presented.

Keywords: Laser Compton scattering γ-ray; nuclear resonance fluorescence; nondestructive assay.

1. Introduction

Nondestructive assay (NDA) of nuclear material based on nuclear resonance fluorescence (NRF) has been proposed for nuclear security and safeguard applications [1,2]. NRF is a process of nuclear excitation by photo absorption and de-excitation by photo emission. It is possible to identify isotopes by detecting characteristic NRF signals. The NRF measurements become more efficient by using a mono-energetic photon beam which can be obtained by laser Compton scattering (LCS) [2]. This has been demonstrated in previous measurements [3,4,5].

Measurement precision of the NRF-based NDA of spent nuclear fuel is affected by background counts from self-activity of spent nuclear fuel and coherent scattering such as Rayleigh, nuclear Thomson, and Delbrück scattering. Since the radiation background of spent fuel exponentially decreases with increasing the energy, NRF measurements at higher energies are preferable to

128

achieve better signal to noise ratio [6]. We have also proposed to measure transitions to the first excited state instead of the ground state to observe NRF signals without overlapping with the coherent scattering background [6]. In this paper, we will show the measurement principle and the estimated counting precision of the measurement system.

2. Measurement principle

Figure 1 shows a schematic view of the measurement principle. Fuel samples are irradiated by LCS γ-rays. NRF γ-rays are detected by a detector array consisting of 24 high purity Ge (HPGe) detectors. Six sets of four HPGe detectors are placed at $\pm120°$, $\pm130°$, and $\pm140°$ with respect to the incident LCS γ-ray beam. The backward angles are selected to decrease the coherent scattering contributions. Lead absorbers are placed between the fuel sample and the HPGe detectors to reduce the background radiation from the sample. The γ-ray beam intensity is measured by a flux monitor. More details are found in Ref. [6].

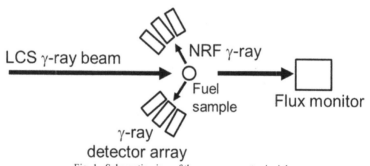

Fig. 1. Schematic view of the measurement principle.

In NRF measurements, transitions to the ground state are usually observed. However, coherent scattering processes such as Rayleigh, nuclear Thomson, and Delbrück scattering scatter beam photons without changing the energy. Therefore, NRF γ-ray peaks overlap with the coherent scattering backgrounds as shown in Fig. 2(a). This gives aproblem when we analyze a small fraction (less than a few %) of the sample. The NRF peak is covered with the coherent scattering backgrounds. In order to analyze NRF peaks, we have therefore proposed to measure transitions to the first excited states. Since the energies of the transitions to the first excited states are 10 to 50 keV smaller than those of the ground state transitions, *i.e.*, energies of the coherent scattering background as shown in Fig. 2(b), the coherent scattering background contributions can be neglectedif we use an incident photon beam with an energy width less than 1% [6].

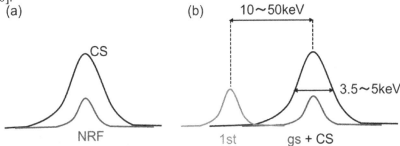

Fig. 2. Illustration of NRF γ-ray and coherent scattering (CS) background peaks. The energy of the transition to the ground state coincides with the coherent scattering background (a). The energy of the transition to the first excited states is 10 to 50 keV apart from the coherent scattering background (b).

3. Counting rates and precision

In the NRF measurements, NRF γ-rays from sample targets are directly detected by γ-ray detectors. In this case, the intensity of the measured NRF γ-rays is proportional to the amount of isotopes of interest. The NRF yield can be calculated by the product of incident γ-ray intensity, integrated cross sections, and target thickness. Assuming the γ-ray intensity of 10^6 1/s/eV, the integrated cross sections of 3 to 100 eV b, branching ratios of 0.33 and 0.67 to the first excited state and the effective target thickness of 0.057 g/cm^2, which is equivalent to the effective thickness of 1% Pu content included in a typical fuel rod, the NRF yield was obtained. Assuming an overall peak detection efficiency of 2.4×10^{-4}, we obtained the NRF count rate of 0.027 to 1.8 count per second (cps).

We have also estimated the effects of the radiation background from a 10-year cooled spent fuel for a typical PWR 17×17 fuel assembly with initial ^{235}U enrichment of 4.1 wt% under burn-up condition of 55GWd/MtUby using the ORIGEN2.2-UPJ computer code [7]. The results show that fission products mainly contribute to the photon intensity at energies lower than 3 MeV while actinides contribute to it at energies higher than 3 MeV. Assuming a detector energy resolution of 0.2%, the radiation background rates summed up for the 24 HPGe detectors at the NRF peak positions are calculated to be 2.0, 1.6×10^{-2}, and 4.6×10^{-4} cps at E=2, 3.5, and 5 MeV, respectively. Figure 3 compares the counting rate of the NRF signals and the radiation backgrounds. The values at

E=3.5 and 5 MeV are smaller than the NRF count rates. The details are described in Ref. [6].

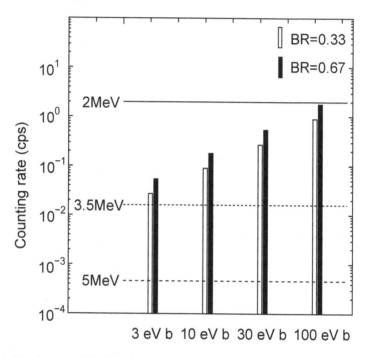

Fig. 3. Counting rates of the NRF signals for transitions to the first excited states assuming the total integral cross sections of 3 to 100 eV b. The values for the branching ratios of 0.33 and 0.67 are shown with the open and filled bars, respectively. Counting rates of the background radiation at 2, 3.5, and 5 MeV are also shown with the solid, dotted, and dashed lines, respectively.

We have estimated the statistical uncertainties by using the equation,

$$\frac{\sigma_S}{S} = \frac{\sqrt{\sigma_T^2 + \sigma_B^2}}{T - B},$$

where S, T, and B are the NRF count, the total count (T=S+B), and the radiation background count, respectively. The corresponding statistical deviations are shown as σ_S, σ_T, and σ_B, respectively. Figure 4 shows statistical uncertainties of the NRF measurements for a 1% Pu content in spent fuel during 4000 seconds measurement time for the NRF peaks located around 3.5 MeV. We assume the total integrated cross sections of 3 to 100 eV b and the branching ratios of 0.33 and 0.67 to the first excited states. For the total integrated cross section of 30 eV b, measurements with statistical uncertainties of 2.2 to 3.2% can be obtained during 4000 seconds measurement time.

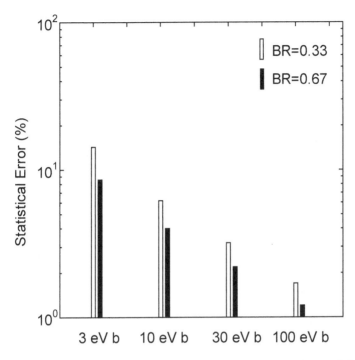

Fig. 4. Statistical uncertainties of the NRF measurements for a 1% Pu content in spent fuel assuming the total integrated cross sections of 3 to 100 eV b and the branching ratios of 0.33 (open bars) and 0.67 (filled bars) to the first excited states during 4000 seconds measurement time for the NRF peaks located around 3.5 MeV.

4. Summary

We have estimated NRF count rates assuming various integrated cross sections. In addition, radiation background counts from spent nuclear fuel were calculated using the burn-up code ORIGEN2.2-UPJ for a typical PWR fuel assembly. When the energy width of the incident photon beam is as small as 0.1%, the coherent scattering contribution can be neglected by measuring transitions to the first excited states, because the energy of the NRF transitions to the first excited state is about 10 to 50 keV lower than that of the coherent scattering. Consequently, it is possible to assay 1% Pu content in spent fuel with 2.2 to 3.2% statistical error during 4000 second measurement time for NRF peaks at E=3.5 MeV with the integrated cross section of 30 eV b.

References

1. W. Bertozzi and R.J. Ledoux, Nucl. Instrum. Methods Phys. Res. B 241, 820 (2005).
2. R. Hajima, *et al.*, J. Nucl. Sci. Tech. 45, 441 (2008).
3. T. Shizuma, *et al.*, Phys. Rev. C 78, 061303(R) (2008).
4. T. Shizuma, *et al.*, Phys. Rev. C 87, 024301 (2013).
5. N. Kikuzawa, *et al.*, Appl. Phys. Exp. 2, 036502 (2009).
6. T. Shizuma, *et al.*, Nucl. Instrum. Methods Phys. Res. A 737, 170 (2014).
7. K. Suyama, *et al.*, A complete package of ORIGEN2 libraries based on JENDL-3.2 and JENDL-3.3.

Improving the Assay of ^{239}Pu in Spent and Melted Fuel Using the Nuclear Resonance Fluorescence Integral Resonance Transmission Method

C.T. Angell*, T. Hayakawa, T. Shizuma and R. Hajima

Quantum Beam Science Directorate,
Japan Atomic Energy Agency,
Tokai, Ibaraki 319-1112, Japan
* angell.christopher@jaea.go.jp
www.jaea.go.jp

B.J. Quiter and B.A. Ludewigt

Acclerator and Fusion Research Division,
Lawrence Berkeley National Laboratory,
Berkeley, CA ZIP/Zone, USA

H. Karwowski and G. Rich

Department of Physics, University of North Carolina at Chapel Hill
Chapel Hill, NC 27599, USA, and
Triangle Universities Nuclear Laboratory,
Durham, NC 27710, USA

Non-destructive assay (NDA) of ^{239}Pu in spent nuclear fuel is possible using the isotope-specific nuclear resonance fluorescence (NRF) integral resonance transmission (IRT) method. The IRT method measures the absorption of photons from a quasi-monoenergetic γ-ray beam due to all resonances in the energy width of the beam. According to calculations the IRT method could greatly improve assay times for ^{239}Pu in nuclear fuel. To demonstrate and verify the IRT method, the IRT signature was first measured in ^{181}Ta, whose nuclear resonant properties are similar to those of ^{239}Pu, and then measured in ^{239}Pu. These measurements were done using the quasi-monoenergetic beam at the High Intensity γ-ray Source (HIγS) in Durham, NC, USA. The IRT signature was observed as a decrease in scattering strength when the same isotope material was placed upstream of the scattering target. The results confirm the validity of the IRT method in both ^{181}Ta and ^{239}Pu.

Keywords: Nuclear material accountancy; spent fuel assay; nuclear resonance fluorescence; safeguards.

1. Introduction

The melting of nuclear fuel in the reactors at Fukushima has introduced a new challenge in accountancy of special nuclear material. Since Japan is a non-nuclear weapon country, the Japanese government has a responsibility to the Japanese people, and to its regional neighbors, to verify that the nuclear material in its possession is used only for peaceful purposes, regardless of the state of the material. To account fully for the Pu in the Fukushima reactors, and verify that a significant amount is not diverted, requires developing advanced, high-precision non-destructive assay (NDA) techniques.

The integral resonance transmission (IRT) method[1,2] is an advanced NDA technique for assaying ^{239}Pu that could significantly reduce measurement time when using transmission nuclear resonance fluorescence (NRF). NRF is a process where the nucleus resonantly absorbs a γ ray (typically of a few MeV) and then de-excites by emitting one or more characteristic γ rays, providing a unique fingerprint of each isotope[3]. The resonance absorption depletes the beam at the resonance energy leaving a notch in the beam energy profile whose depth is proportional to the amount of ^{239}Pu in the fuel. This depletion information can be used to quantitatively assay the Pu in the melted fuel by having a detector system placed in the beam following transmission through the fuel[4]. The transmission detection system consists of a target–termed the transmission target–of the isotope of interest (e.g. ^{239}Pu) and radiation detectors to measure the NRF signal from it. Resonance depletion presents itself as a decrease in the NRF signal from the transmission target for the same amount of incident, non-resonant flux. A transmission measurement overcomes the key issue of the residual radioactivity, unknown shape, and unknown composition of the melted fuel. NRF transmission (self-absorption) measurements using a single resonance have been proposed for detecting smuggled nuclear material[5] and for assaying ^{239}Pu and ^{240}Pu in spent nuclear fuel[6,7].

The IRT method improves the transmission NRF technique by utilizing the integral strength of all resonances excited by a quasi-monoenergetic gamma-ray beam, including weaker states that have not been resolved with current HPGe detectors. The quasi-monoenergetic nature of the beam ensures that the integral signal is due primarily to the NRF signal and not the downscattered beam photons that comprise an irreducible background. The integral signal enables the use of cheaper, faster, more efficient scintillator detectors and uses the additional absorption signature from the many

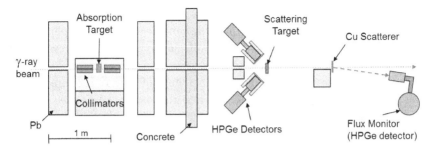

Fig. 1. The experiment on ^{181}Ta consisted of an upstream absorption target placed between the collimators, and a downstream scattering target. The beam moves from left to right in the figure. The beam was collimated by two Pb collimators, and Pb and concrete shielding prevented scattered γ rays from the absorption target to directly enter the detectors. An array of four HPGe detectors, each placed in a Pb annulus shield, measured the scattered γ rays from the scattering target. The flux is measured by monitoring the Compton scattering on the Cu scattering plate using a downstream HPGe detector moved out of the beam by $8°$. In the experiment on ^{239}Pu, the absorption target was placed further up-beam from the collimators. The figure is approximately to scale.

unresolved states, both of which can improve assay sensitivity. This method assumes that there exist unresolved states, as suggested by the discrepancy between expected strength and observed strength of the scissors mode of excitation in odd-mass nuclei[8]. Using these unresolved resonances could improve assay time by over an order of magnitude in certain conditions, according to calculations[1,9]. Measurements are needed to test this basic assumption and demonstrate the validity of the IRT method.

In this paper, we present the first direct evidence for resonance absorption in heavy odd-A nuclei due to unresolved resonances. This is the first test of the IRT method and its assumptions. This evidence comes from transmission measurements of ^{181}Ta and ^{239}Pu. ^{181}Ta was chosen as a simulant material as it is also an odd-mass, odd-Z nucleus, with similar nuclear properties as ^{239}Pu.

2. Measurement

The IRT method was demonstrated by measuring the NRF of ^{181}Ta at two energies, $E_\gamma = 2.28$ and 2.75 MeV, and of ^{239}Pu at $E_\gamma = 2.14$ MeV. Measrurements were done with and without an upstream absorber. The reduction in NRF scattering when an absorber of the same isotope was used is a signature of resonant absorption. The measurements were done at the High Intensity γ-ray Source at Triangle Universities Nuclear Laboratory, in Durham, NC, USA[10]. The beam energy full-width half-maximum was about 4%, or 100 keV, with a typical flux of the order of 10^7 photons/s. The scattering target was placed perpendicular to the beam in a evacuated plastic pipe. The scattered radiation was measured using four 60% relative-efficiency HPGe detectors at an azimuthal angle of 135° to the direction of the beam, two in the horizontal and two in the vertical plane (see Fig. 1). The detectors were shielded by 2 cm of Pb on the sides, and by 5.5 mm Pb and 4 mm Cu in the front. The detectors were placed 14 cm from the target. The γ-ray spectra were measured using a multi-channel analyzer. The flux was determined by measuring the Compton scattering off of a 1-mm-thick Cu plate using a HPGe detector positioned at 8°. The beam energy profile was measured by moving the same detector into the beam, and attenuating the beam by inserting several Cu blocks very far upstream. The beam energy profile was measured approximately every six hours to monitor for possible changes.

For the Ta measurement, the absorption targets were placed between the two Pb collimators (diameters were 19 mm and 22 mm for the up-stream and down-stream collimators, respectively). Three different thicknesses of Ta absorption targets (16, 20, and 30 mm thick) and a Pb absorption target (26 mm thick) were used. The absorption targets were placed 2.5 m up-beam of the scattering target, with shielding in place to prevent scattering from the absorption target directly entering the detectors. The scattering target was a Ta disk 7.4 mm thick with a 31.4 mm diameter. The Ta measurements were taken with no absorbing target for about 4 hrs, and an absorbing target in place for 6-8 hrs.

For the Pu measurement, the absorption target was moved to a position before the collimators. The Pu absorption target (94% ^{239}Pu) was a single 195 g disk, 73 mm in diameter, and 2.7 mm in thickness, encapsulated in several layers of Ta and stainless steel. The target was rotated so that its face was 70° from normal to the beam to increase the effective thickness. An additional Pb shielding wall was built about 20 cm up-beam from the scattering target to further reduce scattering from the absorption target reaching the detectors. The detectors were moved closer–to 11 cm from the target–and the up-beam (down-beam) collimator was changed to 19 mm (16 mm) diameter. The scattering target consisted of eight separate disks of Pu (94% ^{239}Pu, 25 mm diameter), each 1.03 g of Pu encapsulated in Cu (or possibly Fe, neither of which affects the measurement) and then plastic for a total of about 8 g of ^{239}Pu when stacked together. Measurements were done with and without the absorption target in place. The data was collected using a list-mode data acquisition system enabling the use of time-of-flight technique to reduce the radioactive background.

3. Analysis

Raw spectra from the four detectors were first summed together. For the Pu data, timing cuts were also applied reducing the radioactive background by a factor of three to four. The spectra were corrected for pileup. The radioactive background was then subtracted. For comparison, the spectra were normalized correcting for differences in total incident flux and other factors. The normalization method assumed that the beam profile was identical between runs.

3.1. *Pileup Correction*

The pileup correction of the spectra was necessary because of the high detector rates in the experiment, up to 12 kHz per detector, and because the MCA and DAQ system did not have pileup-up rejection. This resulted in a pileup of up to 12%. The spectra were corrected for spectral distortion caused by pileup. Typically, this requires knowing the full spectrum which was known for the Ta measurement[11]. When the full spectrum was not available, as for the Pu measurement when an energy threshold and timing cut were applied, the general pileup function method was developed and used. This method assumes the existence of a single function representing how the counts from a peak are redistributed in energy because of the presence of pileup. This is true over a broad energy region when the spectrum is dominated by an unchanging source of background, concentrated at low energy. This is satisfied in the present case as 98% of the spectrum is due to annihilation photons or Compton scattering below 511 keV. The full spectrum was measured separately (no threshold or timing cuts) to construct the general pileup function. It was then applied to the measured spectrum following timing cuts, with the pileup amount determined from the detector rate. By testing the method with the Ta data, it was determined to introduce a 1% systematic uncertainty.

3.2. *Normalization*

To compare two spectra (such as by taking a ratio), a normalization factor is necessary. The normalization of spectrum 2 (i.e. absorber) to spectrum

138

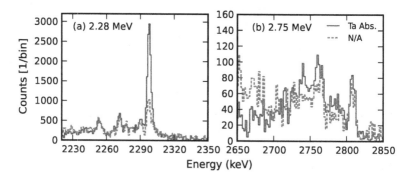

Fig. 2. (a) NRF measurement of ^{181}Ta at $E_\gamma = 2.28$ MeV without (solid line) and with (dashed line) a ^{181}Ta absorber placed upstream of the scattering target with 1 keV/bin. The reduction in the peak heights with the absorber in place is due to resonance absorption. (b) same as (a), but at $E_\gamma = 2.75$ MeV and with 2 keV/bin. Absorption by unresolved states is apparent at 2.78 to 2.80 MeV, demonstrating the principle of the IRT method.

1 (i.e. no absorber) is

$$S = \frac{M_1}{M_2} \frac{L_{m_2}}{L_{m_1}} \frac{L_1}{L_2} \frac{(1 + \rho_2)}{(1 + \rho_1)}, \tag{1}$$

where M_i is the monitor detector counts, L_{m_i} is the livetime of the monitor, L_i is the livetime of the primary detectors, and ρ_i is the pileup amount. The normalization was verified using the Ta data by comparing the ratio of the strong 2.297 MeV NRF peak for no absorber and for a Pb absorber, and was consistent with 1, as expected.

4. Results

In the ^{181}Ta measurement at $E_\gamma = 2.28$ MeV six peaks were resolved, with clear absorption observed for the largest peak. Five transitions have been observed previously using a bremsstrahlung beam [12]; one transition at 2.278 MeV is newly observed here demonstrating the high sensitivity of this measurement (Fig. 2, (a)). Comparing the spectra when using similar thicknesses of Pb and Ta absorption targets demonstrates that the resonance absorption of the strongest transition is due solely to Ta. No absorption was observed when using a Pb absorption target in place of a Ta one.

In the ^{181}Ta measurement at $E_\gamma = 2.75$ MeV, a region at 2.78 to 2.80 MeV was observed with no apparent discrete peaks where γ-rays are absorbed (Fig. 2, (b)). This is evidence of the underlying principle of the IRT

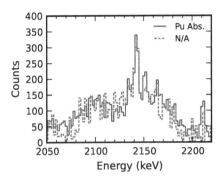

Fig. 3. NRF measurement of ^{239}Pu at $E_\gamma = 2.14$ MeV without (solid line) and with (dashed line) a ^{239}Pu absorber placed upstream of the scattering target with 2 keV/bin. The result shows that the state at 2.15 MeV is the strongest absorbing state in this energy region. There is evidence for scattering by unresolved states at around 2.1 MeV.

method: that there are unresolved states whose resonance absorption can contribute to the total integrated absorption signature. Newly observed resolved transitions at about 2.74 and 2.77 MeV were observed in addition to confirming the 2.76 MeV state and the 2.81 MeV trio of states[12].

The ^{239}Pu measurement at $E_\gamma = 2.14$ MeV demonstrated that the 2.150 MeV state is more strongly absorbing than the 2.143 MeV state (Fig. 3). Bertozzi *et al.*[13], however, indicates the latter having roughly twice the elastic scattering strength of the former. One possible explanation for this is that the absorption strength is larger for the 2.150 MeV state, but that it has strong branching to excited states giving a smaller ground state branching ratio. Additional strength is seen from 2.07 to 2.13 MeV demonstrating the presence of unresolved resonances. Two new states are observed at 2.16 and 2.18 MeV, demonstrating the high sensitivity of this measurement.

5. Conclusions

The reduction in NRF scattering due to self-absorption following transmission in ^{181}Ta was measured at $E_\gamma = 2.27$ and 2.75 MeV, and in ^{239}Pu at $E_\gamma = 2.14$ MeV using quasi-monoenergetic γ-ray beams. The measurements demonstrate the validity of the principle of the IRT method–that there are resonances which are not resolvable, but contribute to the total NRF scattering strength. This was demonstrated for both ^{181}Ta and ^{239}Pu, with ^{181}Ta showing evidence for appreciable absorption by those states. This suggests that using the IRT method may realize a significant improvement to the assay time of the ^{239}Pu in the melted fuel at Fukushima particularly for broad beam sources[1,9].

Acknowledgments

We would like to gratefully acknowledge the contribution of M. Omer, and the efforts of the staff at the High Intensity γ-ray Source in providing high quality beams. We also gratefully acknowledge the help of A. Hunt at Idaho State University for providing one of the Pu targets. Parts of this work was supported by the Office of Nonproliferation and Verification Research and Development, NNSA, US Department of Energy under Contracts No. DE-AC02-05CH11231 and DE-FG02-97ER41041.

References

1. C. Angell, T. Hayakawa, T. Shizuma and R. Hajima, in *Institute of Nuclear Materials Management 53rd Annual Meeting*, July 2012.
2. C. Angell, T. Hayakawa, T. Shizuma and R. Hajima, in *13th Symposium on Advanced Photon Research*, November 2012.
3. T. Hayakawa, N. Kikuzawa, R. Hajima, T. Shizuma, N. Nishimori, M. Fujiwara and M. Seya, Nondestructive assay of plutonium and minor actinide in spent fuel using nuclear resonance fluorescence with laser Compton scattering, *Nucl. Instr. Meth.* **A 621**, 695 (2010).
4. F. Metzger, Resonance fluorescence in nuclei, *Prog. Nucl. Phys.* **7**, 53 (1959).
5. J. Pruet, D. P. McNabb, C. A. Hagmann, F. V. Hartemann and C. P. J. Barty, Detecting clandestine material with nuclear resonance fluorescence, *J. Appl. Phys.* **99**, p. 123102 (2006).
6. B. J. Quiter, B. A. Ludewigt and S. D. Ambers, Assessment of nuclear resonance fluorescence for spent nuclear fuel assay, in *Institute of Nuclear Materials Management 52nd Annual Meeting*, July 2011.
7. B. Quiter, B. Ludewigt, V. Mozin, C. Wilson and S. Korbly, *Nucl. Instr. Meth.* **B 269**, 1130 (2011).
8. U. Kneissl, H. Pitz and A. Zilges, Investigation of nuclear structure by resonance fluorescence scattering, *Prog. Part. Nucl. Phys.* **37**, 349 (1996).
9. B. J. Quiter and T. Laplace and B. A. Ludewigt, *Institute of Nuclear Materials Management 52nd Annual Meeting*, July 2013.
10. H. Weller *et al.*, Research opportunities at the upgraded HIγS facility, *Prog. Part. Nucl. Phys.* **62**, 257 (2009).

11. D. Cano-Ott, J. Tain, A. Gadea, B. Rubio, L. Batist, M. Kany and E. Roechi, Pulse pileup correction of large NaI (Tl) total absorption spectra using the true pulse shape, *Nucl. Instr. Meth. in Phys. Res. A* **430**, 488 (1999).

12. A. Wolpert *et al.*, Low-lying dipole excitations in the heavy, odd-mass nucleus ^{181}Ta, *Phys. Rev. C* **58**, 765 (1998).

13. W. Bertozzi, J. A. Caggiano, W. K. Hensley, M. S. Johnson, S. E. Korbly, R. J. Ledoux, D. P. McNabb, E. B. Norman, W. H. Park and G. A. Warren, Nuclear resonance fluorescence excitations near 2 Mev in ^{235}U and ^{238}Pu, *Phys. Rev. C* **78**, p. 041601 (2008).

Laser Compton Scattering Gamma-Ray Beam Source at NewSUBARU Storage Ring

S. Miyamoto, S. Amano, S. Hashimoto, N. Sakai and A. Koizumi

*Laboratory of Advanced Science and Technology for Industry, University of Hyogo
Kouto, Kamigori, Ako, Hyogo 678-1205, Japan
†E-mail: miyamoto@lasti.u-hyogo.ac.jp*

T. Hashimoto and T. Shizuma

*Quantum Beam Science Directorate, Japan Atomic Energy Agency
Shirane, Shirakata, Tokai, Ibaraki 319-1195, Japan
E-mail: hayakawa.takehito@jaea.go.jp*

H. Utsunomiya, T. Yamagata and H. Akimune

Department of Physics, Konan University, Kobe, 658-8501, Japan

T. Shima

RCNP, Osaka University, Ibaraki, Osaka 567-0047, Japan

D. Li

Institute for Laser Technology, Suita, Osaka 565-0871 Japan

Y. Asano and H. Ohkuma

SPring-8, Sayo, Hyogo 679-5148, Japan

Laser Compton scattering gamma-ray beam source has been developed at the NewSUBARU synchrotron light facility. The available maximum Gamma-ray photon energy is 76 MeV. The flux of quasi-monochromatic gamma-ray photons (for example: 16.7 MeV, $\Delta E/E \sim 5\%$) is more than 10^6 photons/sec using a 35 W Nd:YVO$_4$ laser combined with the 1 GeV storage electron beam with an intensity of 300 mA. We used the electron beams at $E_e = 0.55 \sim 1.47$ GeV for changing the energy of quasi-monochromatic gamma-ray beam. Gamma-ray beams were used for application experiments, a nuclear physics research, a nondestructive inspection of thick material, a generation of positron by pair creation, a magnetic Compton scattering measurements, and a nuclear transmutation.

Keywords: Laser-Compton scattering γ rays; storage ring; photo nuclear reaction.

144

1. Introduction

Laser Compton scattering (LCS) photon sourceis a powerful tool for various scientific investigation and engineering because of its unique characteristics such as quasi-mono energy and highly polarization obtainedby limiting the scattering anglewitha collimator. These features of the LCS gamma-ray are originated from the fact that we use a relativistic electron beam.

In the case of the head-on collisionof a linear polarized photon with a high energy electron, the relativistic electronis forced to vibrate in the direction of the electric field of the laser photon with Lorenz boosted frequency $2\gamma(c/\lambda)$ where $\gamma = E_e/mc^2$ is the Lorenz factor of electrons, E_e is an electron energy, c is the speed of light, m is the electron rest mass and λ is the wavelength of the used laser. The vibrating electrons emit scattering photons with the same frequency and the same polarization of the incoming laser photons witha dipole radiation distribution function of $\sin^2 \phi$(see Fig.1(a)).

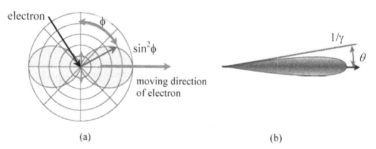

(a)   (b)

Fig. 1. Schematic of Compton scattering of linear polarized photons by relativistic electrons. (a) relativistic electrons scatter the photons in 4π direction with $\sin^2\phi$ distribution. (b) By the Lorenz transformation, the scattered photons concentrate into narrow angle of forward direction.

Observed laser Compton scattering photons at the laboratory frame were the frequency boosted frequency $4\gamma^2(c/\lambda)$ and the Lorenz transformed distribution shown in Fig. 1(b). The photon energy depends on the scattering angle from the beam axis as

$$E_\gamma = \frac{4\gamma^2 E_L}{1 + (\gamma\theta)^2 + 4\gamma E_L/(mc^2)} \qquad (1)$$

where E_γ is a gamma-ray photon energy, $E_L = hc/\lambda$ is a laser photon energy, and h is the Planck constant. Figure 2(a) shows the dependence of gamma-photon energy E_γ on the scattering angle θ calculated by Eq.(1).

Schematic of LCS-γ spectrum in the case of low laser photon energy ($E_L \ll mc^2$) and high electron energy ($E \gg E\gamma$).

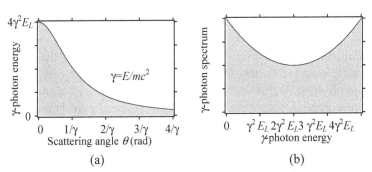

Fig. 2. (a) LCS gamma-photon energy as a function of the scattering angle. (b) Energy spectrum as a function of LCS gamma-ray energies in the case of the head on collision.

A quasi-monochromatic gamma-ray beam is obtained by selecting a solid angle of scattered LCS photons using a collimator. The polarization of scattered LCS photons is theoretically calculated by using the polarization of incident laser. In case of the linearly polarized laser, a high polarization is easily obtained. In general, the polarization of scattered LCS photons in the observation frame is calculated in terms of the rotation of polarization, because the scattered photons are Lorenz transformed from the 4π direction to the forward narrow angle.

We have developed a LCS gamma-ray beam source at the NewSUBARU synchrotron light facility [1-4]. Various experiments using the LCS gamma-ray beam have been carried out for studying nuclear physics [5], positron generation [6], nuclear transmutation [7, 8], and material researches. Recently, the LCS gamma-ray beam-line at New SUBARU was upgraded. An experimental hutch for the photonuclear reaction experiments was newly installed. Thanks to this new experimental hutch, we can increase the gamma-ray beam flux and the photon energy.

2. Gamma-ray Source at NewSUBARU

The NewSUBARU synchrotron light facility consists of an electron storage ring [9] and nine beam lines. The circumference of the storage ring is 118.7 m. Electrons with an energy of 1 GeV are injected from the SPring-8 linac. The Operation energies of the storage ring are 0.5 GeV to 1.5 GeV. In the case of the

146

1.0 GeV operation, the NewSUBARU storage ring is operated with the top-up injection mode at a constant current of 300mA.

The harmonic number of the storage ring (the number of electron bunches) is 198. The acceleration RF frequency is 499.955 MHz.The bunch separation is about 2nsec.The turn around time is 396 nsec. The pulse width of the indivudual electron bunch is about 60 psec [10]. Figure 3 shows the schematic filling pattern of electron bunchesin the storage ring. In the most of the gamma-ray application experiments, this electron filling pattern and the continuous laser injection are used. In the application experiments such as a neutron time-of-flight measurement from photonuclear reaction, we need a good time resolution. In such a case, we use a single bunch operation with a maximum current of 30 mA. In this operation mode, only one electron bunch circulates in the electron storage ring with a 2.525 MHz frequency.

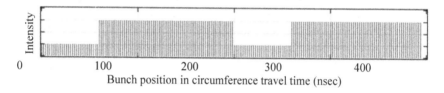

Fig. 3. Filling pattern of the electron bunch current in the NewSUBARU storage ring.

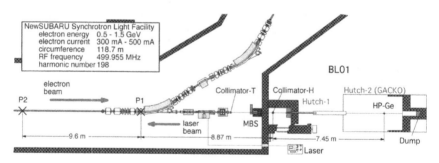

Fig. 4. A layout of the gamma-ray beam-line BL01 at the NewSUBARU synchrotron light facility. Two collision points and two gamma-ray irradiation hutches are shown. The collision point P1 is used for generating the LCS gamma-rays with a CO_2 laser. The P2 is used for the LCS gamma-rays with the Nd(ω) and Nd(2ω) lasers. The LCS gamma-rays at the collision point pass through the final laser mirrors and vacuum window, and are transported to the experimental hatch-1 and -2 through the atmospheric pressure air. Two collimators, collimator-T and –H, are installed in the gamma-ray extraction line to obtain the quasi monochromatic gamma-ray beams. The hutch-2, named GACKO, has been recently installed for collaborative application experiments.

Figure 4 illustrates the experimental setup of the gamma-ray beam line BL01. A new Gamma-ray experimental hutch-2 has recently been installed for gamma-ray beam application experiments. The shielding design of hutch-2 was reported elsewhere [11]. The maximum LCS gamma-ray beam power is 0.33mW which is due to the legal regulation for the radiation safety mainly coming from thin radiation shields. Three kinds of laser systems are used in the experiments; They are (1) a grating-fixed, CW oscillated CO_2 laser with a wavelength of λ =10.5915 μm, (2) a Nd:YVO$_4$ laser with a fundamental wavelength of λ=1064 nm, and (3) a Nd:YVO$_4$ laser with a second harmonic wavelength of λ=532 nm.

Table 1. LCS gamma-ray energies and normalized yield ratesat BL01.

Lasers		Nd(ω)	Nd(2ω)	CO_2
Laser wavelength (nm)		1064	532	10591.5
Electron energy (collimator)		Collimated γ-ray energy range (MeV)		
E_n= 974 MeV	(25 mmφ)	6.0-16.7	12-33	0.6-1.7
	(3 mmφ)	15.8-16.7	31-33	1.6-1.7
E_n= 1470 MeV	(25 mmφ)	7.6-37.6	15-73	0.7-3.9
Electron energy (collimator)		Normalized γ-ray yield rate[a]($\gamma/s/mA/W$)		
E_n= 974 MeV	(25 mmφ)	6000	3000	7200
	(3 mmφ)	260	130	480

Note:[a] Usable flux of gamma-ray is 3000 times of the normalized yield, when we use an electron beam current of 300 mA and a 10 W laser.

Table 1 shows the photon energies and the normalized yield rates of available LCS γ-ray beams at BL01. The LCS gamma-ray energies are calculated with two different electron energy, E_n=974 MeV and E_n=1470 MeV. This electron energies are the nominal value of the NewSUBARU regular operation. The normalized γ-ray yield rate is a round number calculated from a measured value.

3. Energy Calibration of Electron Beams in the Storage Ring [12]

Low-energy LCS γ-ray beams were produced below 2 MeV with a CO_2 laser. Electrons at 1 GeV injected from alinear accelerator into the NewSUBARU storage ring were decelerated to 950 MeV, and were subsequently decelerated down to 550 MeV in steps of 50 MeV. The LCS γ-ray beams were measured with a coaxial (64 mm in diameter × 60 mm in length) high-purity germanium detector (HP-Ge). Figure 5 shows a γ-ray energy spectrum measured with a HP-Ge detector when we generated the LCS gamma-rays using a 800 MeV electron beam and a CO_2 laser with a wavelength of 10.592μm.

148

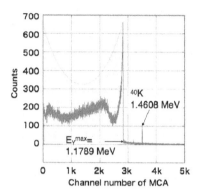

Fig. 5. A measured γ-ray energy spectrum witha HP-Ge detector. A 800 MeV electron beam and aCO₂ laser are used. The diameter of the collimator used is 1 mm. The calculated LCS gamma-ray energy without any collimator is displayed for comparison.

The energy scale of the obtained γ-ray spectra was calibrated by using the standard γ-ray sources, ^{60}Co including the sum peak, ^{133}Ba, ^{137}Cs, and ^{152}Eu and a natural radioactivity ^{40}K. From the calibrated γ-ray spectra, we determined the absolute values of the electron beam energies [12]. The difference ΔE of the calibrated energy E_e from the nominal energy E_n is shown in Fig. 6 as a function of E_n,where the nominal energy is the electron energy used in the control system of the NewSUBARU electron storage ring.

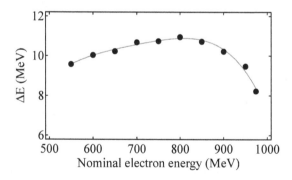

Fig. 6. The difference ΔE of the calibrated energy E_e from the nominal energy E_n of the electron beam, $\Delta E = E_e - E_n$, at the NewSUBARU storage ring.

The fitting result of this data using the fourth-order polynomial function gives

$$\Delta E = -4.6949 \times 10^{-10}(E_n)^4 + 1.3017 \times 10^{-6}(E_n)^3$$
$$- 1.3596 \times 10^{-3}(E_n)^2 + 0.63854(E_n) - 103.94. \qquad (2)$$

Here E_e and E_n are given in MeV. The difference between the nominal and calibrated energies is 10.44 MeV (1.56%) at the nominal energy 667 MeV, which was consistent with the result of the neutron-threshold search for [197]Au.

Acknowledgments

This work was partially supported by JSPS KAKENHI number 25610061, 25400303. The gamma-ray experimental hutch "GACKO" was supported by Konan University.

References

1. S. Miyamoto, Y. Asano, S. Amano, D. Li, K. Imasaki, H. Kinugasa, Y. Shoji, T. Takagi, T. Mochizuki, *Radiation Measurements* **41**, S179 (2007).
2. S. Amano, K. Horikawa, K. Ishihara, S. Miyamoto, T. Hayakawa, T. Shizuma, T. Mochizuki, *Nuclear Instrum. and Methods in Phys. Res.* **A 602**, 337 (2010).
3. K. Horikawa, S. Miyamoto, S. Amano, T. Mochizuki, *Nuclear Instrum. and Methods in Phys. Res.* **A 618**, 209 (2010).
4. T. Kondo, H. Utsunomiya, H. Akimune, T. Yamagata, A. Okamoto, H. Harada, F. Kitatani, T. Shima, K. Horikawa, S. Miyamoto, *Nuclear Instrum. and Methods in Phys. Res.* **A 659**, 462 (2011).
5. T. Hayakawa, S. Miyamoto, Y. Hayashi, K. Kawase, K. Horikawa, S. Chiba, K. Nakanishi, H. Hashimoto, T. Ohta, M. Kando, T. Mochizuki, T. Kajino, and M. Fujiwara, *Phys. Rev. C* 74, 065802 (2006).
6. D. Li, K. Imasaki, S. Miyamoto, K. Horikawa, S. Amano, and T. Mochizuki, *Appl. Phys. Lett.* 94, 091112 (2009).
7. D. Li, K. Imasaki, K. Horikawa, S. Miyamoto, S. Amano, and T. Mochizuki, *J. Nuclear Science and Technology* **46**, 831 (2009).
8. H. Ejiri, T. Shima, S. Miyamoto, K. Horikawa, Y. Kitagawa, Y. Asano, S. Daté, Y. Ohashi, *Journal of Physical Society of Japan* **80**, 094202 (2011).
9. http://www.lasti.u-hyogo.ac.jp/NS-en/facility/ring/para.html
10. Y. Takagi, M. Nakano, K. Arikawa, K. Ishikawa, S. Amano, S. Miyamoto, T. Mochizuki, *J. Synchrotron Radiation* **12**, 826 (2005).

11. Y. Asano, S. Miyamoto, *Proceedings of 12th Int'l Conf. on Radiation Shielding* (ICRS-12), 2–7 September, Nara (2012).

12. H. Utsunomiya, T. Shima, K. Takahisa, D. M. Filipescu, O. Tesileanu, I. Gheorghe, H.-T. Nyhus, T. Renstrøm, Y.-W. Lui, Y. Kitagawa, S. Amano, S. Miyamoto, to be published in *IEEE Transactions on Nuclear Science* **61** no. 3, pp.1252–1258 (2014).

Energy Calibration of Electron and Gamma-Ray Beams at NewSUBARU-GACKO

T. Shima

Research Center for Nuclear Physics, Osaka University,
10-1 Mihogaoka, Ibaraki, Osaka 567-0047, Japan
shima@rcnp.osaka-u.ac.jp

H. Utsunomiya

Department of Physics, Konan University,
8-9-1, Okamoto, Higashinada, Kobe 658-8501, Japan
and Center for Nuclear Study, University of Tokyo,
2-1 Hirosawa, Wako, Saitama 351-0198, Japan
hiro@konan-u.ac.jp

The absolute energies of the electron beam and the laser Compton-scattered (LCS) γ-ray beam provided at the NewSUBARU synchrotron radiation facility were calibrated within the accuracy of the order of 10^{-4} by measuring the LCS γ-ray energies with a energy-calibrated high-purity germanium detector.

Keywords: Energy calibration; laser Compton-scattered γ-ray.

1. Introduction

γ-ray beams with calibrated energies are very useful in various fields of science and engineering. In nuclear physics, quasi-monochromatic γ-ray beams have been used for systematic studies of the resonant excitations of nuclei[1-3]. Standard γ-ray sources have been used to test or calibrate radiation detectors for high-energy photons and electrons. Since γ-rays have high penetrability, they can be used for imaging or radiography for thick materials or materials containing heavy elements[4,5]. Recently new methods by means of resonant photonuclear reactions or excitations have been developed for non-destructive detection and assay of stable as well as radioactive isotopes[6-8]. Development of a new standard γ-ray source with high-intensity and high-quality is, therefore, considered to be worthwhile in wide area of science and engineering.

So far the γ-rays with well-known energies have been obtained from the annihilation of electron-positron pairs, γ-transitions of radioactive isotopes, nuclear radiative capture reactions, and so on. The energy of the annihilation γ-ray has been known[9] with an accuracy of 2.2×10^{-8}. On the other hand, the energies of the nuclear γ-rays are determined with accuracies of $10^{-6} \sim 10^{-4}$, which are limited by the accuracies of the nuclear mass measurements or the crystal spectrometers for high-energy photons[10]. All those γ-rays have fixed energies, almost no polarizations, almost no directivity, and no time-structures, and therefore their applications have been rather limited. On the other hand, γ-rays generated by the laser Compton-backscattering (LCS) process have the following good features [5,11-15];

· considerable intensity (typically $10^6 \sim 10^9$ photons/s),
· small spreads in energy (a few % $\sim$ 0.1%) and angle ($\sim$0.1 mrad),
· variable energy in the range from a few ten keV to a few GeV,
· nearly 100% polarized (linearly or circularly),
· variable pulse frequency from 0 Hz to more than a few GHz,

and so on. The energy-calibrated LCS γ-ray is expected to offer a standard light source in the region beyond the energies of the conventional standard γ-ray sources based on radioactive isotopes.

In the NewSUBARU synchrotron radiation facility of the Laboratory of Advanced Science and Technology for Industry (LASTI), the University of Hyogo, a new experimental hutch named GACKO (GAmma Collaboration hutch of KOnan university) has been constructed, and is now open for the experiments with the LCS γ-ray beam. The absolute energies of the NewSUBARU electron beam and the LCS γ-rays were measured at GACKO, and were calibrated with an accuracy of $10^{-4} \sim 10^{-5}$ by means of the "frequency-scaling" method. In this paper the principle of the energy calibration and the results will be presented.

2. Experimental Method

The calibration of the beam energies were carried out in two steps. First, LCS γ-rays with the energy from 561 keV to 1728 keV were generated via the Compton backscattering of the photons from a CO_2 laser (wavelength $\lambda = 10.5915 \pm 0.0003$Å) and the electron beam with the "nominal" energy E_{nom} from 550 MeV to 974 MeV. The "calibrated" electron beam energy E_{cal} was determined from the measured energy E_γ of the LCS γ-rays by solving Eq. 1 ;

$$E_\gamma = \frac{4\eta^2 \varepsilon_L}{1 + (\eta\theta)^2 + 4\eta\varepsilon_L/(m_e c^2)}, \tag{1}$$

where ε_L is the energy of laser photons, and m_e is the rest mass of electrons. θ is the angle of the scattered γ-ray with respect to the axis of the incident electron beam. η is the Lorentz factor for electrons, given as $\eta = E_{cal}/m_e c^2$. E_γ was measured by using a high-purity germanium (HPGe) detector which was calibrated with standard γ-ray sources and a natural radioactivity. This method is essentially the same as the technique used for calibration of electron beam energies at many facilities in the world[16–21].

Next, the range of the calibration of the HPGe detector was extended to higher energy by means of the frequency-scaling method. In this method the CO_2 laser was replaced with a Nd:YVO$_4$ laser with λ =1064 nm to generate LCS γ-rays in the energy region above 5 MeV. By using those γ-rays, the HPGe detector was calibrated up to $\sim$5.5 MeV.

Finally the light source was again changed to the CO_2 laser, and the LCS γ-rays with the energy from 1.7 MeV to 3.8 MeV were produced by using the electron beam with E_{nom} = 974$\sim$1460 MeV. The absolute energies of those γ-rays were measured with the calibrated HPGe detector, and the real electron energy was determined by using Eq. 1 again.

3. Experimental Setup

The experiment was performed at the GACKO experimental hutch in the BL01 beam line of NewSUBARU. Fig. 1 shows a schematic view of the experimental setup. LCS γ-rays were generated at the collision points P1 and P2 for the CO_2 laser and the Nd:YVO$_4$ laser, respectively. The LCS γ-rays were collimated with two kinematical slits C1 and C2, and were introduced into the Hutch 2 (GACKO). The collimator C1 was placed inside of the shielding of the storage ring, and had an aperture with a diameter of 6 mm. On the other hand, the second collimator C2 had a thickness of 10 cm and an aperture with an diameter of 2 mm. Therefore the scattering angles were defined with C2. The distances of C2 to P1 and P2 were 878 cm and 1847 cm, respectively, and therefore the maximum scattering angles θ_{max} were defined to be 1.14×10^{-4} rad and 5.41×10^{-5} rad for the γ-rays generated with the CO_2 laser and the Nd:YVO$_4$ laser, respectively.

The LCS γ-rays were measured with a HPGe detector (ORTEC GMX45) with a diameter of 64 mm and a length of 60 mm. The axis of the HPGe detector was about 10 mm shifted from the γ-ray axis in order

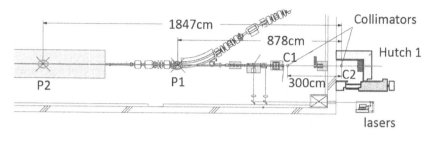

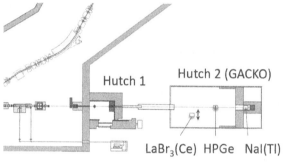

Fig. 1. Schematic view of the experimental setup (top view).

to avoid the γ-rays passing through the central hole of the Ge crystal. At the end of the beam line, a NaI(Tl) detector with a diameter of 6" and a length of 5" was placed as a beam dump and a γ-ray flux monitor.

4. Result

4.1. *Calibration of HPGe detector*

The energy calibration of the HPGe detector was performed by measuring the channels of the full-energy peaks of discrete γ-rays from standard γ-ray sources ^{60}Co including the sum peak, ^{137}Cs, ^{133}Ba and ^{152}Eu, and a natural radioactivity ^{40}K. Fig. 2 shows the relationship between the incident γ-ray energies E_γ and the observed pulse heights. It was found that the observed pulse heights can be fitted with a linear function of E_γ with an overall accuracy of $\sim 6.0 \times 10^{-5}$.

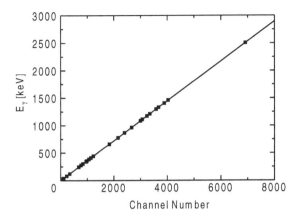

Fig. 2.   Correlation between the γ-ray energy and the pulse height of the HPGe detector.

4.2.   *Calibration of electron beam energy below* $E_{nom} = 974$ *MeV*

The calibrated electron beam energy E_{cal} was determined using Eq .1 from the measurement of the absolute energy of the LCS γ-rays with use of the calibrated HPGe detector. Using a grating-fixed CO_2 laser (INFRARED INSTRUMENTS, IR-10-WS-GF-VP) and the electron beam with the nominal energy E_{nom} between 550 MeV and 974 MeV were used to generate the LCS γ-rays below $\sim$1.8 MeV, and the measured pulse height spectra were compared with the results of the simulation calculation. The simulation calculation was performed by means of the EGS4/PRESTA code[22] which simulated the Compton backscattering process and the interactions of the γ-rays with the detector materials. The production of γ-rays was based on the kinematics and the cross section for Compton backscattering. The cross section in the rest frame of electron was given by the Klein-Nishina formula. In the calculation the calibrated electron beam energy E_{cal}, the electron beam size σ at the collision point, and the effective energy spread δ were assumed as the free parameters, and were determined by fitting the measured spectra with the calculated ones. Here the effect of the electron beam divergence was effectively included by introducing the effective beam size, because both the beam size and the divergence make the same kinematical effect of broadning the energy spread of the LCS γ-rays, and do not affect the maximum energy. Fig. 3 shows an example of the measured and calculated spectra for the LCS γ-rays for $E_{nom} = 850$ MeV. The full-energy peak was well reproduced by the simulation with the parameters of $E_{cal} =$ 860.72 MeV, $\sigma = 1.0$ mm and $\delta = 0.1\%$, where δ consisted of the energy resolution of the HPGe detector 0.077% and the intrinsic energy spread of the electron beam 0.063% .

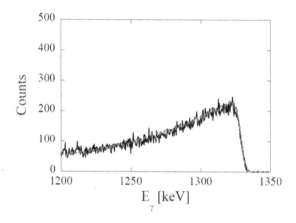

Fig. 3. Pulse height spectra of the LCS γ-ray for $E_{nom} = 850$ MeV. The histogram and the smooth curve indicate the measured spectrum with the calibrated HPGe detector and the calculated one with the EGS4/PRESTA code, respectively.

The accuracy in the determination of E_{cal} was $(1 - 5) \times 10^{-5}$, which was dominated by the statistical fluctuation of the measured pulse height spectra.

The calibration was performed at ten energies in the range of E_{nom} from 550 MeV to 974 MeV, and the energy difference $\Delta E_e \equiv E_{cal} - E_{nom}$ was obtained as a function of E_{nom} as shown by Fig. 4.

4.3. Calibration of electron beam energy above $E_{nom} = 974$ MeV

To calibrate the electron beam energy in the region above $E_{nom} = 974$ MeV, the calibration of the HPGe detector was performed at higer energies. By replacing the CO_2 laser with the Nd:YVO$_4$ laser (SPECTRA-PHYSICS, INAZUMA, $\lambda = 1064$ nm) and using the energy-calibrated electron beam with $E_{nom} = 559.60$ MeV ($E_{cal} = 569.30$ MeV), the LCS γ-rays with the energies of 5528 keV was generated. Using the full-energy peak of 5528 keV γ-ray and its single- and double-escape peaks, the HPGe detector was

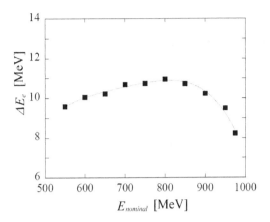

Fig. 4. Energy difference $\Delta E_e = E_{cal} - E_{nom}$ as a function of E_{nom}. The solid curve is an empirical function to approximate the measured data.

calibrated up to about 5.5 MeV as shown by Fig. 5.

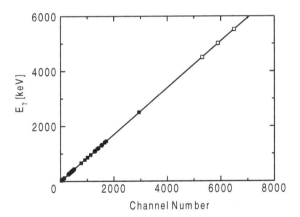

Fig. 5. Correlation between the γ-ray energy and the pulse height of the HPGe detector. The filled squares are the data obtained with standard γ-ray sources and a natural radioactivity. The open squares are the full-energy peak (5528 keV), single-escape peak (5017 keV), and double-escape peak (4506 keV).

It was found that the correlation between E_γ and the channel number could be fitted with a linear function with the relative accuracy of 1.9×10^{-4}.

Then the LCS γ-rays in the energy region between 1.7 MeV and 3.8 MeV were generated using the CO_2 laser and the electron beam with $E_{nom} = 974{\sim}1460$ MeV. From the measured γ-ray energies and Eq. 1, the electron beam energies were calibrated in the energy range of $E_{nom} = 974{\sim}1460$ MeV. The difference ΔE_e between E_{cal} and E_{nom} was obtained as shown by Fig. 6.

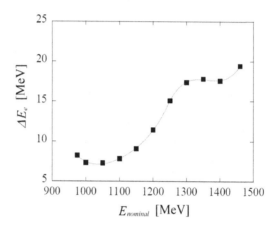

Fig. 6. The energy difference ΔE_e between E_{cal} and E_{nom} in the range of $E_{nom} = 974{-}1460$ MeV.

5. Uncertainty of the Calibration

The uncertainty of the calibration is governed by the uncertainties of three factors included in Eq. 1; the laser wavelength λ, the scattering angle θ, and the γ-ray energy E_γ measured with the HPGe detector.

For the wavelength of the CO_2 laser, we used the P(20) transition with $\lambda = 10.5915$ μm, whose uncertainty and band width are known to be $\pm 3\text{Å}$ [23] and 1.3Å [24], respectively. Therefore the relative uncertainty in the wavelength is quoted as $4.1{\times}10^{-5}$. In the case of the Nd:YVO$_4$ laser, the transition with 1064 nm are doublet with wavelengths of 1064.205 nm and 1064.36 nm [25], and therefore the relative uncertainty in the wavelength is quoted to be about $1.5{\times}10^{-4}$.

The uncertainty of the scattering angle depends on the accuracy in the position of the collimator. It was adjusted to get the maximum flux of the collimated LCS γ-rays, and the error was found to be $\sigma \sim 1$ mm[26], which is comparable to the estimated size of the electron beam at the collision points. Therefore the uncertainties of θ are at most about 2.3×10^{-4} and 1.1×10^{-4} for the CO_2 laser and the Nd:YVO_4 laser, respectively.

The uncertainties in the energy calibrations of the HPGe detector were 6×10^{-5} and $1.9 \sim 10^{-4}$ for the calibrations with only the radioactive isotopes and with the calibrated LCS γ-ray in addition to the radioactive isotopes, respectively.

In summary, including the uncertainties due to the statistical fluctuation in the pulse height spectra of LCS γ-rays, the overall systematic uncertainties are calculated as quadratic sums of each uncertainties, and are evaluated to be 1.4×10^{-4} and 2.5×10^{-4} in the ranges of $E_{nom} = 550-974$ MeV and $974-1460$ MeV, respectively.

6. Summary

The abolute energy of the electron beam of the NewSUBARU synchrotron radiation facility was calibrated with the precision of the order of 10^{-4} in the nominal energy range of $550-1460$ MeV by means of low-energy LCS γ-ray beams produced with a CO_2 laser at the newly constructed γ-ray experimental hutch GACKO. The energy-calibrated electron beams provide the LCS γ-ray beams with precisely determined energies in the ranges of $E_\gamma = 5.6-38.0$ MeV, $11.2-76.0$ MeV and $16.8-114.0$ MeV using photons of the fundamental mode ($\lambda = 1064$ nm), second (532 nm) and third (355 nm) harmonics of the Nd:YVO_4laser, respectively. Those γ-rays will provide a good opportunity for experimental researches to develop nuclear science.

Acknowledgments

We thank Profs. S. Miyamoto, S. Amano, Mr. Y. Kitagawa from the Laboratory of Advanced Science and Technology for Industry, University of Hyogo, Prof. H. Ohgaki from the Institute of Advanced Energy, Kyoto University, and Dr. K. Takahisa from RCNP, Osaka University. We are also grateful to Drs. D.M. Filipescu, O. Tesileanu, I. Gheorghe from the ELI-NP project, Drs. H.T. Nyhus and T. Renstrøm from Department of Physics, University of Oslo, and Dr. Y.-W. Lui from the Cyclotron Institute, Texas A&M University. This work was supported by the Japan Private School Promotion Foundation and partly by the JSPS-FNRS bilateral program.

References

1. B. L. Berman and S. C. Fultz, *Rev. Mod. Phys.* **47**, 713 (1975).
2. H. Ohgaki et al., *Nucl. Phys. A* **649**, 73c (1999).
3. H. Utsunomiya et al., *Phys. Rev. C* **63**, 018801 (2000).
4. H. Toyokawa, H. Ohgaki, and T. Shima, *IEEE Trans. Nucl. Sc.* **49**, 182 (2002).
5. S. Miyamoto et al., *Rad. Meas.* **41**, S179 (2007).
6. H. Ejiri and T. Shima, *Phys. Rev. ST. Acc. Beams* **15**, 024701 (2012).
7. T. Hayakawa, N. Kikuzawa, R. Hajima, T. Shizuma, N. Nishimori, M. Fujiwara, and M. Seya, *Nucl. Instr. and Meth. in Phys. Res. A* **621**, 695 (2010).
8. R. Hajima, T. Hayakawa, N. Kikuzawa, and E. Minehara, *J. Nucl. Sci. Technol.* **45**, 441 (2008).
9. J. Beringer et al. *Phys. Rev. D* **86**, 010001 (2012).
10. C. Doll et al., *J. Res. Natl. Inst. Stand. Technol.* **105**, 167 (2000).
11. R. H. Milburn, *Phys. Rev. Lett.* **10**, 75 (1963).
12. F. R. Arutyunian and V. A. Tumanian, *Phys. Lett.* **4**, 176 (1963).
13. H. Toyokawa, H. Ohgaki, T. Mikado, and K. Yamada, *Rev. Sci. Instr.* **73**, 3358 (2002).
14. V. N. Litvinenko et al., *Phys. Rev. Lett.* **78**, 4569 (1997).
15. T. Nakano et al., *Nucl. Phys. A* **684**, 71c (2001).
16. L. C. Hsu, C.-C. Ching and C.-I. Yu, *Phys. Rev. E* **54**, 5657 (1996).
17. R. Klein, T. Mayer, P. Kuske, R. Thornagel, and G. Ulm, *Nucl. Instr. and Meth. in Phys. Res. A* **384**, 293 (1997).
18. H. Ohgaki, H. Toyokawa, K. Kudo, N. Takeda, and T. Yamazaki, *Nucl. Instr. and Meth. in Phys. Res. A* **455**, 54 (2000).
19. R. Klein et al., *Nucl. Instr. and Meth. in Phys. Res. A* **486**, 545 (2002).
20. K. Chouffani et al., *Phys. Rev. ST Accel. Beams* **9**, 050701 (2006).
21. C. Sun, J. Li, G. Rusev, A. P. Tonchev, and Y. K. Wu, *Phys. Rev. ST Accel. Beams* **12**, 062801 (2009).
22. H. Hirayama, *IEEE Trans. Nucl. Sc.* **40**, 503 (1993).
23. C. K. N. Patel, *Phys. Rev. Lett.* **13**, 617 (1964).
24. R. L. Abrams, *Appl. Phys. Lett.* **25**, 609 (1974).
25. P. B. Lukins, *Characterisation and Calibration of Wavelength Measuring Instruments Based on Grating Spectrometers*, National Measurement Institute, Australian Government (2005).
26. S. Amano et al., *Nucl. Instr. and Meth. in Phys. Res. A* **602**, 337 (2009).

A Paradigm for the Nondestructive Assay of Spent Fuel Assemblies and Similar Large Objects, with Emphasis on the Role of Photon-Based Techniques

Alan Michael Bolind

Integrated Support Center for Nuclear Nonproliferation and Nuclear Security,
Japan Atomic Energy Agency,
3-1-1 Funaishikawa Eki Higashi,
Tokai-mura, Ibaraki-ken, 319-1118, Japan

The practice of nondestructive assay (NDA) of nuclear materials has, until now, been focused primarily (1) on smaller objects (2) with less fissile material and (3) with less self-generated radiation. The transition to the application of NDA to spent fuel assemblies and similar large objects violates these three conditions, thereby bringing the assumptions and paradigm of traditional NDA practice into question for the new applications. In this paper, a new paradigm for these new applications is presented which is based on the fundamental principles of nuclear engineering. It is shown that the NDA of spent fuel assemblies is mostly a three-dimensional problem that requires the integration of three independent NDA measurements in order to achieve a unique and accurate assay. The only NDA techniques that can avoid this requirement are those that analyze signals that are characteristic to specific isotopes (such as those caused by characteristic resonance interactions), and that are neither distorted nor overly attenuated by the other surrounding material. Some photon-based NDA techniques fall into this exceptional category. Such exceptional NDA techniques become essential to employ when assaying large objects that, unlike spent fuel assemblies, do not have a consistent geometry. With this new NDA paradigm, the advanced photon-based NDA techniques can be put into their proper context, and their development can thereby be properly motivated.

1. Introduction

This paper is based upon the author's previously published paper [1] in which he argued that the nondestructive assay (NDA) of spent, enriched-uranium fuel assemblies is a three-dimensional problem, in general. The present paper recasts that argument in summary form and with the purpose of elucidating the unique role that nuclear resonance fluorescence (NRF) plays in the overall art of NDA. This role is NRF's ability to accurately assay large objects that contain a large amount of nuclear material but have an unknown and variable geometry (i.e., the internal, heterogeneous spatial distribution of all material, nuclear and non-nuclear). Large pieces of formerly-molten nuclear-fuel debris from nuclear-reactor meltdowns are such objects.

This paper begins (Section 2) by summarizing the argument for the tri-dimensionality of the important physics of spent fuel assemblies. The next section (Section 3) discusses the main limitation of those NDA techniques that can analyze 'characteristic energy resonances to bypass this typical three-dimensional logic of NDA. This limitation is the attenuation of the radiation signal from the resonances. Section 0 summarizes the limitations described in the previous two sections and discusses NRF's unique ability to transcendthem.

2. Theory (1): The tri-dimensional nature of the logic of the NDA of spent fuel assemblies

The overall logic of spent-fuel-assembly NDA is illustrated in Figure 1. Moving through the figure from left to right, it is seen that the first vector space consists of the quantities that are determined by NDA measurements, such as count rates or ratios of count rates. The second vector space consists of the physical properties of the spent fuel assembly at the time of the NDA measurement. Examples of such physical properties are the number of primary neutrons (per second), the neutron leakage multiplication, the neutron fate probability for capture, and the total gamma-ray activity. The third vector space is what this author has abbreviated as the "BIC set of parameters" and consists of the burnup (BU) of the fuel assembly, its initial enrichment (IE), and its cooling time (CT) since being discharged from the reactor. The values of these parameters come from the historical records of the fuel assembly. The fourth vector space (on the top) consists of the isotopic (nuclide) composition of the spent fuel assembly. The elemental plutonium content is of particular interest to safeguards. The fifth vector space (on the bottom) consists of the contribution of the spent fuel assembly to the physical properties of a collection of spent fuel assemblies, such as in a storage cask. This vector space is of particular interest to the burnup-credit community; they care most about the reactivity that the fuel assembly adds to the collection. This fifth vector space has been mentioned here for completeness but will not be discussed further in this paper.

Most nondestructive assay operates by means of correlations among the vector spaces. Moving again from left to right, it is seen that multiple NDA measurements can be made to determine key physical properties of the spent fuel assembly. The correlation can be either analytical or empirical. The physical properties can then be correlated empirically to the BIC set. This correlation is made by determining the physical properties of a suite of spent fuel assemblies that spans the domain of the BIC-set vector space. Lastly, the isotopic content of each member of the suite of fuel assemblies is determined either by performing destructive analysis on it or by doing a burnup simulation of it. By means of all these correlations, the measured values from NDA can be interpreted as implying specific and accurate values of the isotopic content.

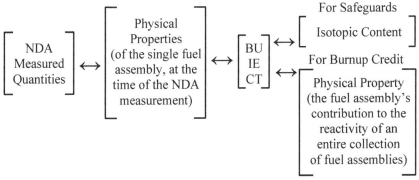

Figure 1: The overall logic of the NDA of spent fuel assemblies

Until now, the vector space of the physical properties has usually been overlooked, and instead the NDA measurements have been correlated directly to the BIC set. That is, the NDA measurements have been made on the suite of spent fuel assemblies, but no effort has been made to determine the physical properties of the assemblies. This shortcut method has two problems. Firstly, the physical properties are the variables that actually govern the NDA measurements, not the BIC parameters. The physical properties are the present characteristics of the fuel assembly and can be repeatedly verified, whereas the BIC parameters represent the past characteristics of the assembly, which cannot be repeated. Secondly, it is difficult to logically combine the data from multiple, different NDA measurements when each measurement has its own correlation to the BIC set. In such a case, the way in which one NDA measurement relates to another is not obvious. Nevertheless, it is essential to combine multiple NDA measurements to achieve an accurate assay result, as will be demonstrated below. Therefore, the best practice is to include the physical-properties vector space in the NDA logic.

Each of the vector spaces in the logic in Figure 1 is three-dimensional (3-D). The dimensionality of the BIC set is three dimensional by choice, but the dimensionality of the physical properties and the isotopic content is inherent to the fuel assembly. This dimensionality comes from the fact that the dominant physics and isotopes of the fuel assembly (for NDA purposes) are its neutronic physics and isotopes, which are each 3-D. The neutronic physics dominates because the way in which the fuel assembly becomes spent (by nuclear burning through fission chain reactions) and the reasons for caring about the fuel assembly after it has been spent (e.g., the fissile quality of its uranium and plutonium) are all related to its neutronic physics.

The tri-dimensionality of the vector spaces of the neutronic physics and isotopic content can be seen in the one-speed neutron diffusion equation as applied to the NDA [2, 3]:

$$S + (\bar{v} - 1)\Sigma_f \phi - \Sigma_{a,capture}\phi - (-D\nabla^2 \phi) = \frac{1}{v}\frac{\partial \phi}{\partial t} \qquad (1)$$

The first term, S, is the generation of primary neutrons, which is dominated by the spontaneous fission of ^{240}Pu, ^{242}Cm, and ^{244}Cm. The second term is the generation of secondary neutrons from induced fission, primarily in the fissile isotopes, ^{235}U, ^{239}Pu, and ^{241}Pu. The third term is the loss of neutrons by capture, in the set of the transuranic nuclides and the set of the seventeen most important fission products for neutron capture (the sixteen listed in Ref. [4] plus ^{155}Eu). The fourth term is the diffusion of neutrons throughout the fuel assembly and out of it. However, this term is not relevant for NDA of spent fuel assemblies because it does not change significantly with the BIC set. It is governed by the geometry of the fuel assembly (which also includes boundary conditions for the diffusion equation), by inelastic neutron scattering in ^{238}U, and by elastic scattering in water, none of which change with the burning of the fuel (i.e., with the BIC set). The fifth term, on the other side of the equation, is the change in the neutron flux over time, but it is either zero for a steady-state NDA measurement or is the output of a time-dependent measurement (e.g., a characteristic die-away time, τ). Of course, all terms change with neutron flux (ϕ) and with position, but though the magnitude of the flux changes with the BIC set, the spatial distribution stays practically constant, as evidenced by the typically homogenous transverse burnup profile [5, 6] and the constant neutron leakage probability [7]. Therefore, it is seen that only three terms in Eq. 1—corresponding to three physical processes and to three groups of isotopes — change with the BIC set and are therefore relevant to NDA.

The gamma-ray physics follows the neutronic physics, in general. Measurements of prompt and delayed gamma-rays and of gamma-rays from fission products correspond to the second term of Eq. 1. Measurements of neutron-capture gamma-rays, such as in the techniques of neutron resonance capture analysis (NRCA) and prompt gamma activation analysis (PGAA), correspond to the third term. Background overwhelms other gamma-rays.

3. Theory (2): The attenuation of signals from characteristic energy resonances

The fact that Eq. 1 does not account for neutron energy (since it is only one-speed) does not invalidate the general conclusion of tri-dimensionality. Again, the constancy of the elastic and inelastic neutron scattering with the BIC set implies that the energy characteristics of the fuel assembly as a whole must also be constant with the BIC set and therefore irrelevant to NDA. Of course, some neutron NDA techniques, such as self-interrogation neutron resonance densitometry (SINRD), lead slowing-down spectroscopy (LSDS), and neutron resonance transmission analysis (NRTA), operate by examining the neutron energy spectrum for characteristic resonance peaks or chasms from specific nuclides. Such techniques are governed, though, by the quantities of those specific nuclides in the fuel assembly while being merely limited because of the attenuationby the assembly as a whole. In short, such neutron NDA techniques directly detect specific nuclides, rather than following Eq. 1 and the general logic of Figure 1. In a similar way, the gamma-ray NDA techniques that analyze characteristic gamma-rays of specific energies also directly detect specific nuclides and thereby bypass the logic of Figure 1 to some extent. Therefore, the tri-dimensionality of the vector spaces in Figure 1 still holds as valid.

This ability of some neutron and photon NDA techniques to bypass the general NDA logic of Figure 1 is valuable, but almost all of them are simultaneously crippled by the attenuation of the energy peak or chasm (i.e., the NDA signal) by the material intervening between the point of generation of the radiation and the detection of it in the detector. Regarding neutrons, a mere one centimeter of water attenuates a characteristic resonance peak or chasm by 99% [8]. For LSDS, which analyzes only dry fuel assemblies, attenuation is manifested as the very difficult problem of self-shielding. Regarding gamma-rays, the vast majority of the energy-dependent, passive gamma-ray signal from a used fuel assembly comes from its outer three rows of fuel pins [9], because the outer pins attenuate the signal from the inner pins. The higher energy of the gamma-rays of the high-energy delayed gamma-ray spectroscopy (HEDGS) NDA technique does reduce their attenuation, but the fission that generates them makes the technique dependent on Eq. 1 and the logic of Figure 1. Therefore, although NDA techniques that analyze characteristic energy resonances provide unique and independent nuclide information, most of them are unsuitable for the assay of a used fuel assembly because they cannot detect a sufficient signal from the inner fuel pins.

4. Results and Discussion: Summary of the two fundamental problems of the NDA of spent fuel assemblies and other large objects, and discussion of how NRF overcomes them

The previous two sections (2 & 3) have presented the two main obstacles facing the NDA of spent fuel assemblies and other large objects. Almost all NDA techniques either (1) require a three-dimensional correlation through the BIC set in accordance with Figure 1 or (2) are prevented, by attenuation, from assaying the inner portions of the assembly or object. In the case of spent fuel assemblies, the fact that the geometry of the assembly is known and constant with the BIC set means that a three-dimensional correlation through the BIC set is possible. Thus, the accurate NDA of spent fuel assemblies is theoretically possible. In the case of other large objects with unknown and variable geometry, though, such a correlation is theoretically impossible, since the diffusion term in Eq. 1 is no longer constant or known. Furthermore, since attenuation prevents a direct assay without a correlation, the accurate NDA of such large objects is fundamentally impossible — with *almost* all NDA techniques.

The only NDA techniques that are able to assay large objects with unknown, variable geometry are those that do not require a 3-D correlation and that can accurately account for the attenuation of the signal. Such techniques are those that analyze characteristic resonance signals that are imparted to a one-dimensional beam of radiation as it is transmitted through the object. The characteristic resonances make such techniques independent of the 3-D correlation, and calibration measurements without the object in the beam allow the attenuation of the signal by the object to be determined accurately as long as at least some of the signal is able to pass all the way through the object. The neutron NDA technique that meets these requirements is NRTA, and the corresponding gamma-ray NDA technique is NRF. Thus, only these two NDA techniques have the capability to assay accurately large objects with unknown and variable geometry, such as melted-nuclear-fuel debris. This conclusion is illustrated by Table 1, which uses the above rationale to evaluate the sixteen NDA techniques that were examined by the Next Generation Safeguards Initiative [10–12] and that serve as a representative sample of the span of possible NDA techniques. Since NRTA is limited to assaying fuel assemblies that are dry and that have at most 12 pins in the beam [13], NRF, which has neither limitation, is the superior NDA technique.

Table 1: Evaluation of sixteen NDA techniques regarding their ability to assay large objects with unknown and variable geometry (N = no; Y = yes)

The 14 NGSI NDA Techniques	Does it directly measure a character-istic signal?	Does it account for attenuation?	Therefore, can it assay large items with unknown geometry?
Total Neutron (TN)	N	–	N
Passive Neutron Albedo Reactivity analysis (PNAR)	N	–	N
^{252}Cf Interrogation with Prompt Neutron detection (CIPN)	N	–	N
Differential Die-Away analysis (DDA)	N	–	N
Assembly Interrogation with Prompt Neutron detection (AIPN)	N	–	N
Delayed Neutron counting (DN)	N	–	N
Neutron Multiplicity (NM)	N	–	N
Differential Die-away Self-Interrogation (DDSI)	N	–	N
Total Gamma-ray counting (TG)	N	–	N
Passive Gamma-ray spectroscopy (PG)	Y	N	N
Delayed Gamma-ray (DG)	N	N	N
Self-Interrogation Neutron Resonance Densitometry (SINRD)	Maybe	N	N
X-Ray Fluorescence (XRF)	Y	N	N
Lead Slowing-Down Spectroscopy (LSDS)	Y	N (self-shielding)	N
Neutron Resonance Transmission Analysis (NRTA)	Y	Y	**Y**
Nuclear Resonance Fluorescence (NRF)	Y	Y	**Y**

5. Conclusion

This paper has discussed the two main limitations of the NDA of spent fuel assemblies and other large objects that contain much nuclear material, and it has shown how NRF transcends them. The first limitation is that most NDA techniques require a 3-D correlation through the BIC set because the neutronic physics is both dominant and three-dimensional. The second limitation is the severe and unpredictable attenuation of the characteristic energy-resonance signals that are used by most of the remaining NDA techniques to detect nuclides directly and thereby avoid needing 3-D correlations. The only two NDA techniques that can avoid both limitations are NRTA (neutrons) and NRF (gamma-rays), because they analyze characteristic energy resonances and yet their one-dimensional beam geometries allow attenuation to be taken into account. These two techniques are therefore the only ones that can accurately assay large objects with unknown and variable geometry, such as melted-nuclear-fuel debris, since 3-D correlations are unavailable for such objects. Furthermore, since NRF is even less affected by attenuation than NRTA is, NRF is the superior NDA technique.

References

[1] A. M. Bolind, *Ann. Nucl. Energy* **66**, 31 (2014).

[2] J. J. Duderstadt and L. J. Hamilton, *Nuclear Reactor Analysis*, John Wiley & Sons, Inc., New York, 331 (1976).

[3] J. R. Lamarsh and A. J. Baratta, *Introduction to Nuclear Engineering*, Prentice Hall, Upper Saddle River, New Jersey (2001).

[4] C. V. Parks, M. D. DeHart, and J. C. Wagner, *Review and Prioritization of Technical Issues Related to Burnup Credit for LWR Fuel*, ORNL/TM-1999/303 and NUREG/CR-6665, Oak Ridge National Laboratory, Oak Ridge, Tennessee (2000).

[5] H. Ezure, *J. Nucl. Sci. Technol. (Tokyo, Jpn.)* **27**, 461(1990).

[6] J. D. Galloway, H. R. Trellue, M. L. Fensin, and B. L. Broadhead, *JNMM* **40**, 25 (2012).

[7] Y. Nauchi, T. Kameyama, A. Sasahara (Central Research Institute of Electric Power Industry, Japan), *Numerical simulations of neutrons transport in spent LWR fuel assemblies for burn-up credit application techniques*, Track 1–27, International Conference on Nuclear Criticality (ICNC 2011), 19–23 September 2011, Edinburgh, Scotland, IBC Energy (2011).

[8] A. M. Bolind and M. Seya, *A collection and distillation of the important physics for understanding, interpreting, and evaluating neutron-based*

non-destructive assay techniques for nuclear fuel, 53rd Annual Meeting of the Institute of Nuclear Materials Management (2012).

[9] J. R. Phillips and G. E. Bosler, *Calculated response contributions of gamma rays emitted from fuel pins in an irradiated PWR fuel assembly*, LA-9837-MS, Los Alamos National Laboratory, Los Alamos, New Mexico (1983).

[10] Stephen Croft, *et al.*, *A Technical Review of Non-Destructive Assay Research for the Characterization of Spent Nuclear Fuel Assemblies Being Conducted Under the US DOE NGSI*, #11544, WM2011 conference, 27 February, 2011, Phoenix, Arizona, WM Symposia (2011).

[11] W. S. Charlton and M. A. Humphrey, *JNMM* **40**, 12 (2012).

[12] M. A. Humphrey, K. D. Veal, and S. J. Tobin, *JNMM* **40**, 6 (2012).

[13] J. W. Sterbentz and D. L. Chichester, *Further evaluation of the Neutron Resonance Transmission Analysis (NRTA) technique for assaying plutonium in spent fuel*, "INL/EXT-11-23391, Idaho National Laboratory, Idaho Falls, Idaho (2011).

Medium Modification of α Cluster Size in ^{6}Li

T. Yamagata[1,*], H. Akimune[1], S. Nakayama[2], T. Shima[3] and S. Miyamoto[4]

[1]*Department of Physics, Konan University, Kobe 658-8501, Japan*

*yamagata@center.konan-u.ac.jp

[2]*Department of Physics, University of Tokushima, Tokushima 770-8502, Japan*

[3]*Research Center for Nuclear Physics, Osaka University, Osaka 567-0047, Japan*

[4]*Laboratory of Advanced Science & Technology for Industry (LASTI), University of Hyogo, Hyogo 678-0242, Japan*

The giant dipole resonance (GDR) in ^{6}Li was investigated via the ^{6}LI(γ, xn) reaction, where $x = 1$, 2 or 3 at an incident energy range of E_γ=5-55 MeV. The (γ, n) cross section was the most dominant cross section among them. The GDR in ^{6}Li was found to consist of two components at E_x= 11 MeV and 33 MeV. The component at E_x=11 MeV seems to be the intrinsic GDR in ^{6}Li. The other at E_x=33 MeV is inferred to be the GDR due to the α cluster excitation in ^{6}Li, based on the comparison with the results in light ion reactions. The GDR in free ^{4}He is known to locate at E_x= 26 MeV. However, the GDR excitation energy due to the α cluster excitation in ^{6}Li is found to be higher than that of the ^{4}He. This fact suggests that the size of the α cluster in ^{6}Li is smaller than that of ^{4}He due to the nuclear medium effect.

Keywords: α cluster-excitation; nuclear medium effect; size of α cluster.

1. Introduction

Presence of α clusters in light nuclei is a common phenomenon. If the α clusters in a nuclear system are weakly bound and recognized as spatially localized subsystems, an intrinsic excitation of the α clusters itself are expected. Indeed, Cost et al. suggested a possible excitation of α cluster in ^{6}Li in a photonuclear reaction [1]. They have observed two resonances at excitation energy of E_x=11.5 and 26 MeV in ^{6}Li and have interpreted that the resonance at 11.5 MeV is the isovector giant dipole resonance (GDR) in ^{6}Li and the resonance at 26 MeV is the excitation of αcluster, namely, the GDR of ^{4}He in ^{6}Li because the GDR in free ^{4}He is known to locate at E_x=26 MeV. Bazhanov et al. also reported peaks at E_x ~ 15 and ~30 MeV [2]. However, no such evidence for the GDR in the α cluster has been obtained in other ^{6}Li(γ, n) reactions [3].

Recently, in the nuclear reactions of (p, p'), (^{7}Li,^{7}Be) and (^{3}He,t) on the ^{6,7}Li targets the dipole resonances have been observed at Q ~ -30 MeV, i.e. excitation energies much higher than those of the GDR's in the target nuclei [4-9]. Based on the comparison of the resonance energies, widths, excitation cross sections and the charged-particle decay modes of these dipole resonances with those of the GDR and the spin dipole resonance (SDR) in^{4}He, it was concluded that the resonance observed at Q ~ -30 MeV were assignedto be the GDR's and SDR's of the α cluster embedded in ^{6,7}Li and their analogs in ^{6,7}He and ^{6,7}Be [4-9].Thus, it is commonly recognized that there exists the excitation of the α cluster.

In the nuclear reactions both GDR and SDR are excited. On the other hand the GDR is selectively excitedin the photonuclear reaction. The properties of the α cluster can be different from those of the free ^{4}He due to nuclear medium effect. Therefore we have a chance to detect the difference between in the α cluster and in free ^{4}He by observing the GDR via photonuclear reactions. In the present work, we investigated the ^{6}Li(γ, n) reaction by using a quasi-mono-energy γ-rays in an energy range of E_γ=5-55MeV.

2. Experimental setup

Theγ-rays were generated via the Compton backscattering of laser photons with relativistic electrons in the synchrotron, NewSUBARU at the LASTI, University of Hyogo [10]. The Nd-laser photons of the fundamental (λ=1064 nm) or the second harmonic (532 nm) modes at 20 kHz were incident into a straight section of NewSUBARU and were collided with electrons. Theelectron energy was varied from 550 to 1300 MeV. The generated γ-rays were collimated by using a

10 cm-thick lead with a 2mm aperture located 14.5m downstream of the collision point. The full-width at half-maximum (FWHM) of the γ-ray energy was evaluated with a Monte-Carlo simulation of EGS4. Resulting values were typically 4.5%.

A target located at the center of a neutron detector in the hutch, GACKO, 7.5m downstream of the collimator, and was bombarded by γ-rays. They-rays passing through the target were detected by using a 6"-diameter and 5"-thick NaI spectrometer. The number of the incident γ-rays was evaluated from this data. The target used was an enriched metallic ^{6}Li (96%). We also used a D_2O (99%) target to calibrate the detection efficiency of the neutron detector. Each target material was contained in a 9 cm-length and 8 mm inner-diameter capsule with 30μm Mylar windows at an entrance and an exit.

As shown in Fig. 1, we used a 4π-type neutron detector. The neutrons were detected by using totally fourty-one ^{3}He-gas proportional detectors with a length of 40 cm. They were embedded parallel to the γ-rays into a 60 cm cubic polyethylene modulator, and located at concentric positions, as like four rings, from the target at 6 cm (3 detectors, which we call as A-ring), at 7.5 cm (9 as B-ring), at 10 cm (11 as C-ring) and, at 12 cm (12 as D-ring).

Each signal from the proportional detector was fed into an amplifier/discrimination (amp/discri) module. The logic signals from the amp/discri in each ring were summed and fed into scalar modules. These signals were also used as the trigger signals of CAMMAC. Signals from the amp/discri were analyzed by 41ch CAMMAC ADC's within a 10μs gate period. The coincidence events of two and three ADC's were used to evaluate the $(\gamma, 2n)$ and $(\gamma, 3n)$ cross sections. In order to subtract the background events due to the electron beams, data were taken in every 100ms with 80 ms laser-on and 20 ms laser-off. The events at laser off time were assumed to be the background, and background counts were subtracted from the counts at laser on time.

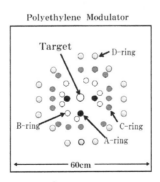

Fig. 1. A front view of the neutron detector. ^{3}He proportional detectors are arranged at positions as A, B, C and D rings.

3. Detection efficiency of neutrons

Since the most excited states in ^{6}Li decay into many particles, as shown in Fig. 2, determination of the energy as well as detection efficiency of the neutrons are very important in the present study. Energy dependence of detection-efficiency of the neutron detector was calculated by using a Monte-Carlo simulation with a code MCNP. On the other hand, we measured the detection efficiency to be 42% by using a ^{252}Cf neutron source and normalized the calculated efficiency by the measured efficiency at 2.4 MeV, which is the mean energy of neutrons from ^{252}Cf.

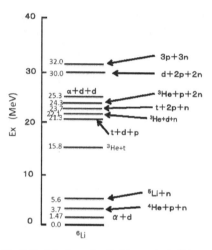

Fig. 2. Particle decay threshold energies in ^{6}Li.

The mean energy of detected neutrons at each incident γ-ray energy was derived from the measured "ring ratios" which were the N_A/N_T, N_B/N_T, etc. the ratio of counts N_A in the ring A to the total counts N_T, and N_B in the ring B to N_T and so on, where $N_T=N_A+N_B+N_C+N_D$. These ring ratios were sensitive to the neutron energy. At first, we measured the ring ratios as a function of a neutron energy by using the $D(\gamma, n)$ reaction with a D_2O target at a range of the γ-ray energy from 5MeV to the threshold energy of the $^{16}O(\gamma, n)$ reaction, i.e. 15MeV. In this reaction the energy of neutrons is given to be $E_n = (E_\gamma - E_D)/2$, where E_D is the deuteron binding energy of 2.225 MeV. In the ^{6}Li(γ, n) reaction, the neutron energy was determined by comparing the ring ratios measured in the $D(\gamma, n)$ reactions. The energies of the neutron and detection efficiency were 0.5

MeV and 47%, respectively, at E_γ= 5 MeV and 5MeV and 35%, respectively, at E_γ= 55 MeV.

4. Results and conclusions

The observed cross sections for the ⁶Li (γ, n_{Total}), (γ, $2n$) and (γ, $3n$) are shown in Fig. 3, where $\sigma(\gamma, n_{Total}) = \sigma(\gamma, n) + \sigma(\gamma, 2n) + \sigma(\gamma, 3n)$. The $\sigma(\gamma, 2n)$ and $\sigma(\gamma, 3n)$are negligibly smaller than the $\sigma(\gamma, n)$. The GDR in ⁶Li is found to consist of two peaks at E_x = 11 MeV and 33 MeV. This result is consistent with that obtained in the light ion reactions [4-9]. The peak at E_x =33 MeV corresponds to the excitation of the GDR in αclusters.

In the photonuclear reactions, Berman et al. observed only the peak at E_x = 12 MeV [3], while Costa *et al.* and Bazhanov *et al.* observed cross sections [1,2] similar to those reported in the present work, though the magnitude of the cross sections were different from those in the present work. In the present work, the energy integral of the cross sections up to E_x = 55 MeV is found to be 59 mbMeV. This exhausts 65% of the TRK Sum Rule and about twice of the value of 27.8 mbMeVreported by Berman [3].

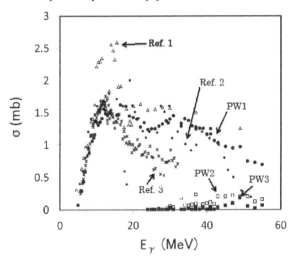

Fig. 3. The⁶Li(γ,n) cross sections. PW1, PW2 and PW3 are the present data for $\sigma(\gamma, n_{Total})$ = $\sigma(\gamma, n) + \sigma(\gamma, 2n) + \sigma(\gamma, 3n)$, $\sigma(\gamma, 2n)$ and $\sigma(\gamma, 3n)$, respectively. Errors for PW1, PW2 and PW3 are estimated to be 7%, 15% and 50%, respectively.

The excitation energy of the GDR in⁴He is known to be E_x ~ 26 MeV. The presently observed excitation energy of the α cluster in ⁶Li, *i.e.* 33 MeV is about 7 MeV higher than that of ⁴He. Even we subtract the separation energy of the αparticle in ^{6}Li, which is 1.4 MeV, the excitation energy of theα clusters is higher than that in free ^{4}He by about 5 MeV. The excitation energy of the GDR is inversely proportional to the size of the system. Therefore, the present result suggests that the size of theα cluster in ^{6}Li nucleus is smaller than that of a free ^{4}He by about 20%.

Acknowledgments

We thank Profs. H. Utsunomiya and M. Fujiwara for their valuable discussions and suggestions. This work was performed by using NewSUBARU-GACKO (Gamma Collaboration Hutch of Konan University). This work was supported by Japan Society for the Promotion of Science, Grant-in Aid for Scientific Research (C) with contact No. 23540349.

References

1. S. Costa *et al.*, Phys. Lett. 4, 308 (1963).
2. E. B. Bazhanov *et al.*, Nucl. Phys. 68, 191 (1965).
3. B. L. Berman and S. C. Fultz, Rev. Mod. Phys. 47, 713 (1975).
4. S. Nakayama *et al.*, Phys. Rev. Lett. 87, 122502 (2001).
5. T. Yamagata *et al.*, Phys. Rev. C 69, 044313 (2004).
6. T. Yamagata *et al.*, Phys. Rev. C 74, 014309 (2006).
7. S. Nakayama *et al.*, Phys. Rev. C 76, 021305(R) (2007).
8. T. Yamagata *et al.*, Phys. Rev. C 77, 021303(R) (2008).
9. S. Nakayama *et al.*, Phys. Rev. C 78, 014303 (2008).
10. K. Horikawa, S. Miyamoto and T. Mochizuki, Nucl. Instruments and Methods in Phys. Res. A 618, 209 (2010).

Noninvasive Reactor Imaging Using Cosmic-Ray Muons[*]

H. Miyadera[†], K. Fujita, Y. Karino, N. Kume,
K. Nakayama, Y. Sano, T. Sugita and K. Yoshioka

Toshiba Corporation,
Yokohama, Kanagawa, Japan
[†]*haruo.miyadera@toshiba.co.jp*

C. L. Morris[‡], J. D. Bacon, K. N. Borozdin and J. O. Perry
P-25, Los Alamos National Laboratory
Los Alamos, NM 87545, USA
[‡]*cmorris@lanl.gov*

S. Mizokami, Y. Otsuka and D. Yamada

Tokyo Electric Power Company
Tokyo, Japan

Cosmic-ray-muon imaging is proposed to assess the damages to the Fukushima Daiichi reactors. Simulation studies showed capability of muon imaging to reveal the core conditions. The muon-imaging technique was demonstrated at Toshiba Nuclear Critical Assembly, where the uranium-dioxide fuel assembly was imaged with 3-cm spatial resolution after 1 month of measurement.

Keywords: Cosmic-ray muon; muon imaging; Fukushima Daiichi.

1. Introduction

Three reactors at Fukushima Daiichi are believed to have considerably damaged due to simultaneous loss of total AC and DC power as well as cooling for an extended period caused by a 15-m tsunami on March 11, 2011. Two weeks later, the three reactors were stabilized by water cooling, and official cold-shutdown condition was announced by the end of 2011. The Japanese government has

---

[*] This work is partially supported by Toshiba Corporation, Tokyo Electric Power Company and by the Laboratory Directed Research and Development Program of Los Alamos National Laboratory.

drawn mid- and long-term roadmap towards decommissioning,[1] and the debris removal from the reactors is scheduled to start as early as 2020. In the case of Unit 2, it was estimated by Modular Accident Analysis Program(MAAP) that 57% of the fuel could have fallen from the damaged part of the reactor pressure vessel (RPV) in the worst case,[2] however, the portion of remained fuel had a wide range and the actual status of the core is unknown. Realistic information of the core damage and the locations of the fuel debris is needed to start the decommissioning work. A camera endoscope entered inside the primary container in January, 2012, but could not find any fuel debris. Still more, little is known for the inside of the RPVwhere the radiation level is dangerously high.

Cosmic-ray muon radiography was proposed to reveal the status of fissile material inside the RPV from outside the reactor building. Cosmic-ray muons are created when highly energetic particles collide with the Earth's upper atmosphere. Initially, pions and kaons are created which decay into muons, and at ground level, the muon flux is 10^4 /m^2/min. Muon imaging was invented in the 1950s and, since then, it was used to study mine overburden,[3] an Egyptian pyramid,[4] a temple gate,[5] volcanoes,[6-9] a blast furnace[10] and caverns.[11]These observation was performed by measuring directions and fluxes of transmitted muons with single-sided muon trackers. Here, we call the traditional muon-imaging method "transmission method". In the case of the transmission method, changes in the rate of muon fluxcaused by objects stopping some of the muons were measured, and the spatial resolutions of the reconstructed images suffered from muon scatterings along the muon pass .[12]

A more sensitive technique, muon scattering method, was invented at Los Alamos National Laboratory. [13]The technique uses multiple Coulomb scattering of muons, and was developed to detect shielded packages of nuclear materials in a background of normal cargo. [14]The scattering method uses at least two muon trackers to measure incoming and outgoing tracks of individual muons, and images the region contained within the acceptance of the tracker pair. In the case of scattering method, locations of the major muon scatterings can be spotted, because both the incoming and outgoing tracks of individual muons are measured and are combined. The scattering method is selective to high-Z dense materials, its spatial resolution is an order of magnitude better than that of transmission method, and has been used by several groups.[15-17] A Gaussian width of multiple Coulomb scattering in various material is given by the following formula ,[18]

$$\vartheta_0 = \frac{13.6}{\beta c p} \sqrt{\frac{x}{X_0}} \left[1 + 0.038 \ln\left(\frac{x}{X_0}\right) \right] \qquad (1)$$

Where θ_0 is the width of scattering angle, βc and p are the velocity and momentum of the incident muon, and x and X_0 are the thickness and radiation length of the scattering medium. Regarding materials used in reactors, the radiation lengths of water, concrete, steel and uranium are 39.3, 11.6, 1.76 and 0.317 cm, respectively.

Though both the scattering and transmission methods were proposed to image the reactors of Fukushima Daiichi, Monte Carlo simulations showed that the transmission method has poor identification capability to image fuels or debris through thick shielding of the reactors while scattering method delivers reliable images to assess the damage to the reactor cores.[19]

2. Feasibility Studies of Fukushima Daiichi Muon Imaging

2.1. *Monte Carlo simulation*

To test feasibility of the reactor imaging with the scattering method, Monte Carlo simulations were performed, and Unit 2 of Fukushima Daiichi was modeled with GEANT4 as shown in Fig. 1.[20]For the simulation study, we implemented a cosmic-ray-muon generator to reproduce correct energy spectrum of muons for different zenith angles in good agreement with known measurements.[21]Some of our Monte Carlo results are published in our previous paper where muon-scattering method provided quantitative assessment of the intact fraction of the core and spherical debris as small as 20-cm radius was distinguished with measurement span of a few months.[22]

Unit 2 was chosen for the first measurement, because its uranium fuel is believed to be mostly contained inside the RPV. The latest plan to image Unit 2 is to installa muon tracker of 7×7-m^2 area on the 2nd floor of turbine building (operation floor), and another 7×7-m^2 tracker in front of the reactor building. Muon trackers for the Unit-2 measurement are currently under development at Toshiba Corporation. The muon tracker in front of the reactor building will be contained in a steel shielding of 7.5-cm thickness or combined-steel-and-lead shielding that gives equivalent attenuation to gamma rays. The radiation level at the proposed installation spot is several hundreads of µSv/h and should be reduced to ~50 µSv/h for stable operation of the muon tracker. On the other hand, the radiation level on the 2nd floor of the Unit-2 turbine building is below 20 µSv/h, therefore, the muon tracker can be operated without any shielding.

180

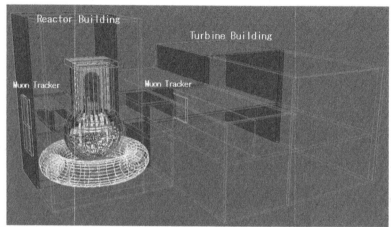

Fig. 1. GEANT4 model of Fukushima Daiichi Unit 2.[21]The model included all the major structures such as reactor core, RPV, major water pipes, reactor container, and reactor and turbine buildings.

The muon tracker in front of the reactor building will be initially installed at ~10-m above the ground level, and later will be moved to the ground level. In the former configuration, muons from the west (direction of reactor building) can image the original core region while the latter configuration can image bottom of the RPV by using muons from the east (direction of turbine building). Results of GEANT4 simulation are shown in Fig. 2 which corresponds to total of 180-days measurements: 90-days measurement with muon tracker in front of the reactor building at 7-m above ground, and another 90-day measurement on the ground.

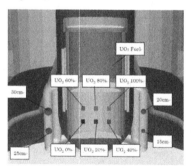

Fig. 2. GEANT4 model of Unit 2 near the core region (left). The core had an average density of 4.3 g/cm^3 over the volume and consisted of uranium dioxide (60.5%), zirconium (22%), stainless steel (2%) and water (15.5%). [23]Spherical debris of 15-, 20-, 25- and 30-cm radii are put inside the water pipes, which has density of 8.2 g/cm^3 and consisted of uranium (70%), zirconium (14%), oxygen (13%) and stainless steel (3%).[24] Also, cubic mixture of uranium-dioxide and zirconium are put in below the core.

2.2. Technical demonstration

As a precursor to the measurement of Fukushima Daiichi, a technical demonstration of reactor imaging with muon scattering method was carried out at Toshiba Nuclear Critical Assembly (NCA) in Kawasaki, Japan, in the summer of 2013.[25] The NCA is a research reactor used for experiments on nuclearfuel. Though the similar technical demonstrations of the reactor imaging was performed with a mockup reactor at Los Alamos and with a research reactor at University of New Mexico (AGN-201),[26] the NCA demonstration was the world's first muon imaging of a reactor with actual uranium-dioxide fuel assembly. The demonstration was carried out for one month (July 19 to August 19, 2013). Roughly speaking, the demonstration was 1/10 scale of the Fukushima Daiichi measurement.

A configuration and results of the NCA demonstration are shown in Fig. 3. A pair of muon trackers, Mini Muon Tracker (MMT), was installed at Toshiba NCA, which was developed at Los Alamos National Laboratory. The MMT has a detection area of 1.2×1.2 m^2, and consists of 576 drift-tube detectors (96 for each x-y plane; 3 x-y planes per supermodule; 2 supermodules) that have dimension of 5-cm diameter and 1.2-m length. The drift tubes are filled with a gas mixture of Ar (50%), CF_4 (44%), and C_2H_6 (6%), and have gold-plated-tungsten anode wires of 30-μm diameters, which are operated at 2.55 kV relative to the ground (tube walls). For the fuel assembly, a 40-cm diameter uranium-dioxide core was assembled with 424 fuel rods which have an average density of 3.5 g/cm^3. The core had a hollow cylinder of 20-cm diameter, and was similar in shape to the melted core of Three Mile Island. Two fuel bundles of 4cm×4cm cross section were installed by the core as well. Concrete and steel bricks with thicknesses of 20 cm and 12 cm, respectively, were put in as obstacles.

The accumulated muon event that penetrated the two muon trackers was 2,800 ~ 2,900/h throughout the measurement. The fuel assembly was identified after several hours of measurement, and the cavity of the core was identified after a few days.[25] Scattering method stores data in three dimension, and images can be reconstructed at various depth between the two trackers. As shown in Fig. 3, all the features of the NCA including the void of the core, the 4cm×4cm fuel bundles, the concrete and steel obstacles were identified at the corresponding depth. The analysis was based on point of closest approach, where the scattered angle is plotted at the intersection of incoming and outgoing muon tracks.[27]

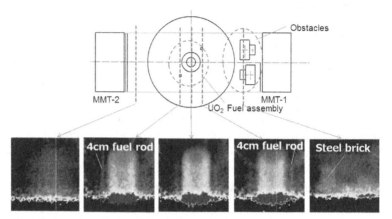

Fig. 3. Configuration of the Toshiba NCA demo (top). Images reconstructed at various depth between the two muon trackers (bottom).

3. Operation of Muon Trackers under High Radiation Environment

One of the major challenge for muon-imaging Fukushima Daiichi reactors is the high gamma-ray radiation of the site, which is mostly from ^{137}Cs and ^{134}Cs. Two problems are caused forthe muon truckers consisting with drift tubes by the gamma rays: high event rate and gas aging of drift tubes. The former can be solved by hardware pattern-recognition logic taking advantage of the difference of muon and gamma-ray events as schematically shown in Fig. 4. The muon trackers currently under development for Fukushima Daiichi equip a FPGAlogic capable of removing 99% of the gamma-ray events to allow the trackers to be operated under radiation environment of up to 50 μSv/h.

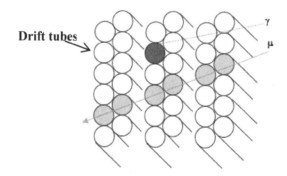

Fig. 4. Difference between muon and gamma-ray events to a muon tracker consisting of drift tubes. Muons penetrates all the layers of drift tubes whereas gamma rays only trigger a single tube.

The aging of drift tube under the radiation environment of Fukushima Daiichi is a long-term effect. It was found that hydrocarbon gases polymerize and create solid whisker on the anode wire of drift tubes,[28] since electrons near the anode wire have sufficient energy to brake hydrocarbon molecules. To estimate the life time of drift tubes at Fukushima Daiichi, we radiated drift tubes with ^{60}Co source as shown in Fig. 5. The following drift gases were tested: a) drift gas containing hydrocarbon (Ar:CF$_4$:C$_2$H$_6$ = 50:44:6); b) non-hydrocarbon drift gas (Ar:CO2:N2 = 96:3:1).[29] The result is shown in Fig. 6, where detection efficiencies of the tubes were plotted as a function of radiation dose. The hydrocarbon gas showed steep decline in detection efficiency with less than 3-month equivalent dose to operation at Fukushima Daiichi, while non-hydrocarbon gas showed no change even after 4-year equivalent dose. Based on the result, the non-hydrocarbon gas was selected for the drift gas of the muon trackers.

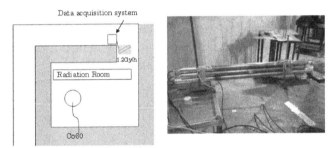

Fig. 5. Experimental set up of drift tube test (left). Drift tubes were radiated with ^{60}Co source.

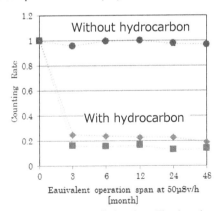

Fig. 6. Detection rates of drift tubes and radiation dose. The detection rates were measured with a checking source. Drift tubes without hydrocarbon contents showed no aging after 4-year equivalent dose to Fukushima Daiichi operation.

4. Conclusion and Future Remarks

Monte Carlo simulations and a technical demonstration with a research reactor were performed as feasibility studies of muon scattering imaging of the Fukushima Daiichi reactors. Muons are strongly deflected by high-Z materials such as uranium, which can be used to reveal the status of uranium fuel in the Fukushima Daiichi reactors. A detector radiation test was performed, and drift gas that could stand more than a few years of operation at Fukushima Daiichi was selected. Currently, the measuring of Unit 2 is aimed at starting in 2015.

Acknowledgments

We acknowledge staffs at Toshiba NCA for their support during the reactor imaging demonstration.

References

1. Japanese Ministry of Economy Trade and Industry, Mid-and-long-term roadmap towards the decommissioning of TEPCO's Fukushima Daiichi nuclear power station Unit 1-4 (2013).
2. Tokyo Electric Power Company, MAAP analysis and Core concrete reaction, November 30, 2011.
3. E. P. George, Cosmic rays measure overburden of tunnel, Commonwealth Engineer, 455 (July 1, 1955).
4. L. W. Alvarez et al., Search for Hidden Chambers in the Pyramids, Science 167, 832, (1970).
5. S. Minato, Feasibility of cosmic-ray radiography: A case study of a temple gate as a testpiece, Mater. Eval. 46, 1468 (1988).
6. K. Nagamine, M. Iwasaki, K. Shimomura, and K. Ishida, Method of probing inner-structure of geophysical substance with the horizontal cosmic-ray muons andpossible application to volcanic eruption prediction, Nucl. Instr. and Meth. A 356, 585 (1995).
7. H. K. M. Tanaka, K. Nagamine, S. N. Nakamura, and K. Ishida, Radiographic measurements of the internal structure of Mt. West Iwatewith near-horizontal cosmic-ray muons and future developments, Nucl. Instr. and Meth. A 555, 164 (2005).
8. J. Marteau et al., Muons tomography applied to geosciences and volcanology, Nucl. Instr. and Meth. A 695, 23 (2012).
9. Felix Fehr, Density imaging of volcanos with atmospheric muons, J. Phys. Conf. Ser. 375, 052019, (2012).

10. K. Nagamine et al., Probing the inner structure of blast furnaces by cosmic-ray muon radiography, Proc. Jpn. Acad. B 81, 257 (2005).

11. G. G. Barnaföldi et al., Portable cosmic muon telescope for environmental applications, Nucl. Instr. and Meth. A 689, 60 (2012).

12. N. Lesparre et al., Geophysical muon imaging: feasibility and limits, Geophys. J. Int. 183, 1348 (2010).

13. K. N. Borozdin et al., Surveillance: Radiographic imaging with cosmic-ray muons, Nature 422, 277 (2003).

14. C. L. Morris et al., Tomographic Imaging with Cosmic Ray Muons, Science & Glorbal Security 16, 37 (2008).

15. S. Pesente et al., First results on material identification and imaging with a large-volume muon tomography prototype, Nucl. Instr. and Meth. A 604, 738 (2009).

16. V. Anghel et al., Prototype testing and algorithm development for the Cosmic Ray Inspection and Passive Tomography (CRIPT) project, Nucl. Sci. Symp. and Med. Imag. Conf. 2011 IEEE, 959.

17. A. A. Borisov et al., A Muon Tomograph setup with a 3×3 m^2 area of overlapping, Instr. and Exp. Tech. 55, 151 (2012).

18. G. R. Lynch and O. I. Dahl, Approximations to multiple Coulomb scattering, Nucl. Instr. Meth. B 58, 6 (1991).

19. K. Borozdin et al., Cosmic Ray Radiography of the Damaged Cores of the Fukushima Reactors, Phys. Rev. Lett. 109, 152501 (2012).

20. S. Agostinelli et al., Geant4—a simulation toolkit, Nucl. Instr. and Meth. A 506, 250 (2003).

21. H. Jokisch et al., Cosmic-ray muon spectrum up to 1 TeV at 75° zenith angle, Phys. Rev. D 19, 1368 (1979).

22. H. Miyadera et al., Imaging Fukushima Daiichi reactors with muons, AIP Advances. 3, 052133 (2013).

23. US Department of Energy, Integrated Data Base Report 1996, DOE/RW-0006-Rev. 13, 1–8.

24. D. W. Akers, S. M. Jensen, and B. K. Schuetz, Examination of relocated fuel debris adjacent to the lower head of the TMI-2 reactor vessel, U.S. Nuclear Regulatory Commission, NUREG/CR-6195 (1994).

25. C.L. Morris et al., Analysis of muon radiography of the Toshiba nuclear critical assembly, Appl. Phys. Lett. 104, 024110 (2014).

26. J. Perry et al., Imaging a nuclear reactor using cosmic ray muons, J. Appl. Phys. 113, 184909 (2013).

27. L. J. Schultz et al., Image reconstruction and material Z discrimination via cosmic ray muon radiography, Nucl. Instr. and Meth. A 519, 687 (2004).
28. J. Kadyk et al., Wire chamber aging, Nucl. Instr. and Meth. A 300, 436 (1991).
29. Nicola Tyler, Studies on Linear and Fast Drift Gases for ATLAS MDT Chambers, Master Thesis, Ludwig-Maxililians Universität (2011).

Compton Radiation for Nuclear Waste Management and Transmutation

E. Bulyak

NSC KIPT, Kharkov, Ukraine
bulyak@kipt.kharkov.ua

J. Urakawa

KEK, Tsukuba, Ibaraki, Japan
junji.urakawa@kek.jp

Compton inverse radiation is emitted in the process of backscattering of the laser pulses off the relativistic electrons. This radiation possesses high spectral density and high energy of photons—in hard x-ray up to gamma-ray energy range—with moderate electron energies (hundreds of MeV up to 1 GeV) due to short wavelength of the laser radiation. The Compton radiation is well collimated: emitting within a narrow cone along the electron beam. A distinct property of the Compton inverse radiation is a steep high-energy cutoff of the spectrum and the maximal intensity just below the cutoff. The Compton sources can attain: spectral density up to 10^{14} gammas/(s 0.1% bandwidth) in MeV range of energies, and spectral brightness up to 10^{20} gammas/(s mm^{2}mr^2 0.1% bw).

Applicability of Compton sources for nuclear waste management and detection of radioisotopes and fissionable nuclides are discussed in the report. Also application limits of Compton gamma sources for transmutation of radioactive isotopes are estimated. A recently proposed subtracting method, in which two sets of data obtained by irradiating the object by the Compton beams with slightly different maximal energies are compared, will enhance resolution of detection radioactive elements at the 'atomic' (hundreds of keV) and the 'nuclear' (a few MeV) photon energies.

Keywords: Compton radiation; nuclear waste management.

1. Introduction

Sources of Compton radiation, in which photons of intense laser pulse are scattered off from relativistic electrons, are able to produce bright x-ray beams with narrow bandwidth. The process of Compton scattering can be treated as two-particle elastic scattering: one of the particle is represented by the electron, another one by the photon. Since the laser photon has negligibly small energy as compared with electron's, the recoil of electron is negligibly small—the photon is scattered off within a narrow cone along electron's trajectory.

Such sources have a substantial potential for applications in different areas of medicine, biology, physics, etc. This potential is emerged not only from its brightness, but tunability and 'quasimonochromaticity' of the spectrum.

The report highlights spectral properties of Compton sources and presents a potential application of these sources for x-ray angiography, which possesses a substantial advantage over the conventional methods.

2. Spectrum of Compton inverse radiation

In general, the spectrum is dependent on the following factors: the collimator opening angle, the energy spread of electrons and photons and the angular spread of their trajectories at the interaction point (IP). In practical cases, the angular spread of photon 'trajectories' produces least impact on the spectrum, as it can be seen from the kinematics: Its effect is proportional to the spread squared because IP is usually set up at the laser waist, while the electron spread impact is γ^2 times larger. Also the spread of electron energy within the bunch usually much higher than that of photons. Therefore below we will consider effects of the electron bunch phase volume—spreads of energy and trajectories—upon the collimated spectrum of scattered off laser quanta.

The spectrum is determined by two relations,

- kinematic dependence of the energy of a scattered off quantum, E_{x} upon the crossing angle ϕ between the laser photon and the electron, energy of both the electron and the photon, and the scattering angle ψ, and
- the differential cross section—dependence of probability of scattering the photon at angle ψ.

Within the small-angle approximation, $\psi \ll 1$ (the crossing angle $\phi = 0$ corresponds to the head-on collision), neglected the electron recoil, these relations read[1]:

$$E_{\mathrm{x}} = \frac{2\gamma^2(1 + \cos\phi)E_{\mathrm{las}}}{1 + \gamma^2\psi^2} \; ; \qquad \mathrm{d}\sigma_{\mathrm{c}} = 8\pi r_0^2 \frac{\psi\gamma^2\left(1 + \gamma^4\psi^4\right)}{\left(1 + \gamma^2\psi^2\right)^4}\mathrm{d}\psi \; , \qquad (1)$$

where $\gamma = E_{\mathrm{e}}/m_{\mathrm{e}}c^2$ is the Lorentz factor of the electron, r_0 is the classical electron radius.

2.1. *Spectral–angular density*

In order to facilitate spectral evaluation we introduce the spectral–angular density. It is defined as:

$$\nu = \frac{1}{\sigma_c} \frac{\partial^2 \sigma_c}{\partial \psi \partial \epsilon} , \qquad (2)$$

with $\sigma_c = 8\pi r_0^2/3$ being the Compton (Thomson) cross section.

Within the small-angle approximation for the scattering angles, the spectral–angular density (2) reads:

$$\nu\left(x, \zeta\right) = \frac{3}{2} \frac{x\left(1 + x^4\right)}{\left(1 + x^2\right)^4} \frac{\delta\left(x - \sqrt{1/\zeta - 1}\right)}{\zeta^2 \sqrt{1/\zeta - 1}} , \qquad (3)$$

where

$$\zeta = \frac{E_x}{E_x^{\max}} = \frac{1}{1 + \gamma^2 \psi^2} \; ; \; x = \gamma \psi ,$$

with δ being Dirac's δ–function.

The formula (3) represents *a spectral–angular density of photons emitted by single electron with respect to its trajectory and energy.* The energy spectrum connected to (3) may be obtained by integration over the angle:

$$G(\zeta) = \frac{3}{2} \left[1 - 2\zeta \left(1 - \zeta\right)\right] , \qquad 0 < \zeta \leq 1 . \qquad (4)$$

The ideal spectrum is symmetric around $\zeta = 1/2$, its minimum at this energy is half of the maximum at $\zeta = 1$. The average spectral density of Compton spectrum—number of photons within 0.1 % bandwidth—is approximately equal to $Y/1000$, where Y is the total number of scattered off photons (per second).

2.2. *Collimated spectrum of real electron bunches*

Convolution of (3) with distribution of the electrons over trajectories and energy within the bunch produces the spectral–angular density of radiation emitted by a source.

Integration of the source density over the range of opening angles (ψ_i, ψ_f) with respect to the bunch central trajectory (we suppose a ring–slit collimator, Fig. 1) produces the spectrum of radiation passed through the collimator:

$$G(\zeta, x_i, x_f) \equiv \int_{x_i}^{x_f} \nu\left(x, \zeta\right) \, dx . \qquad (5)$$

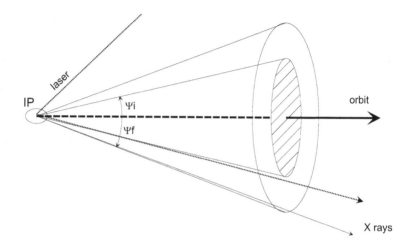

Fig. 1.   Scheme of the ring–slit collimation.

The collimated energy spectrum with account for a small energy spread reads:

$$G_\gamma(\zeta, x_i, x_f) = \frac{3}{4} \left[1 - 2\zeta(1 - \zeta)\right] \left[\mathrm{Erf}(\eta_i) - \mathrm{Erf}(\eta_f)\right] , \qquad (6)$$

where $\eta_{i,f} = \left(1/\zeta - 1 - x_{i,f}^2\right)/2\sqrt{2}s_\gamma$, $\mathrm{Erf}(z)$ is the error integral function, $s_\gamma \equiv \sigma_\gamma/\gamma \ll 1$ is the reduced rms energy spread.

The collimated spectrum with account for the angular spread reads:

$$G_\psi(\zeta, x_i, x_f) = \frac{3\left[\zeta^2 + \left(1 - \zeta^2\right)^2\right]}{2\sqrt{2\pi}s_\psi}$$
$$\times \left[\mathrm{erf}\left(\eta_f^-\right) + \mathrm{erf}\left(\eta_f^+\right) - \mathrm{erf}\left(\eta_i^-\right) - \mathrm{erf}\left(\eta_i^+\right)\right] , \qquad (7)$$

with $\eta_{i,f}^\pm = \left(x_{i,f} \pm \sqrt{1/\zeta - 1}\right)/\sqrt{2}s_\psi$, $s_\psi \equiv \sigma_\psi\gamma$ being the angular rms spread of trajectories at IP multiplied by the Lorentz factor of electrons.

The collimated partial spectra for different ψ_i and the same $\psi_f - \psi_i = 0.02$ are presented in Fig.2 for realistic spreads. These figures show that at each angle with respect to the bunch center-of-weight trajectory, the spectrum of Compton radiation is different. The angular spread of electron trajectories is predominant.

The collimated x/gamma-ray spectra generated by Compton sources have are follow properties which are essential for applications:

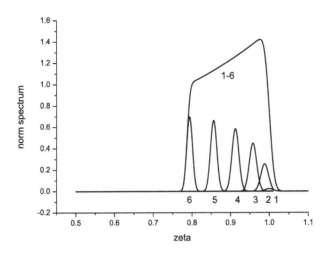

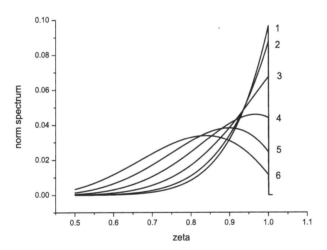

Fig. 2. Spectra at different angles: Curves $1, 2, \ldots, 6$ correspond to $\psi_i = 0, 0.1, \ldots, 0.5$. Top: $s_\phi = 0, s_\gamma = 0.5\,\%$, bottom: $s_\phi = 0.247, s_\gamma = 0$.

- The spectrum width is limited by the high energy cutoff, no photons with higher energy.
- The spectrum has the maximum close to the high-energy cutoff.
- High-energy·cutoff of the spectrum is steep.

- The collimator opening angle should not be more narrow than the (approximately doubled) angular spread of electron trajectories at IP. More narrow collimation will reduce the spectral maximum but not the width.

3. Compton sources and applications

Major figure of merit for the Compton sources is yield Y—total number of quanta generated in unit of time. The yield is proportional to the average current of electron beam and to energy contained in the laser pulse[2]. The maximal yield per electron–pass for a tomorrow laser system (single resonator) is about 0.01. For real electron bunches it is expected to be 2–3 orders lower.

The beams of relativistic electrons are expected to provide by three types of the accelerators as follow:

- Maximal beam current, $I_{max} \leq 5$ A, is provided by storage rings (SR). Expected spectral density up to 10^{14} photon/(s 0.1% bw) if array of resonators used. Also the SR is most efficient since the relativistic electrons may circulate along the orbit for a long time, up to one hour. Only small fraction of the total energy is required for the recovery during circulation. Other side of the coin is least tunability of the device.
- Energy recovery linacs (ERL) are able to provide smaller average current $I_{max} \leq 100$ mA and therefore the spectral density. ERL is less effective than SR since smaller fraction of the beam energy can be recovered. The energy spread in ERL beams is much smaller than that of SR's but it is not essential for the applications considered here.
- The linear accelerators (linacs) are least effective since the beam must be dumped after single pass through the laser pulse without recovery. Also, the average current is $I_{max} \leq 10$ mA. The linacs possess maximal flexibility: fast tune of the energy, sweep of the beam direction, etc. Their use is limited to generation of x-rays, $E_x \leq (100 \ldots 150)$ keV.

3.1. Applications

3.1.1. Nuclear resonance fluorescence

The most demanded application for the nonproliferation and the nuclear waste management is the nuclear resonance fluorescence (NRF), that requires high spectral density of gammas in the energy range (1–5) MeV because of small cross section, of order $\sigma_{nrf} = (1–30)$ barn eV.

Radiation of the Compton source is able to yield up to $\sim 10^{10}$ NRF photons per second from 1 mol/sq.cm density material, that is in good excess of that required for the nuclear forensics and waste elemental analysis.

3.1.2. X-ray K-edge imaging

This method is able to detect heavy elements on background of lighter elements. Necessary are steep high-energy cutoff of the spectrum and quick enough change of electron's energy (energy of the cutoff). The method proposed for angiography[3] is able to reveal small density of an element in x-ray absorbtion images made with two maximal energies of the Compton spectrum, lower and higher the K-edge of the element.

For example, as it can be seen from Fig. 3, to detect uranium on the lead background it is necessary to get x-ray images with the maximal Compton spectra energy 90 keV and 120 keV, respectively.

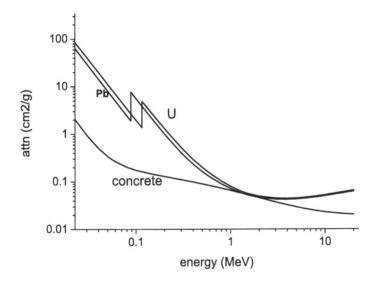

Fig. 3.   Attenuation factors for concrete, lead and uranium[4].

3.1.3. Transmutation.

An attractive way of transmutation is exciting the giant dipole resonance—wide, and with big cross section—for fission a radionuclide. Energy of this resonance, approximately $E_{\text{gdr}} \approx 80 \, \text{MeV} \times A^{-1/3}$ with A being the atomic number of an element. This energy lays within range of Compton gamma sources that yield maximum brightness among others.

Cross section of giant dipole resonance is $\sigma_{\text{gdr}} \approx 3 \times 10^{-27} A \, \text{cm}^2$. For example, the most intense Compton gamma source (based on the storage ring) produces 10^{15} gammas/s. Being 'focused' on $1 \, \text{mm}^2$ area, it produces flux 10^{17} gammas/$(\text{s}\,\text{cm}^2)$. Life time of nuclei, τ, with photo disintegration cross section σ and gamma flux n_γ is

$$\tau^{-1} = \sigma n_\gamma \, .$$

Accordingly, for $^{137}_{55}\text{Cs}$ (with the cross section $\sigma = 4.11 \times 10^{-25} \, \text{cm}^2$ (0.411 barn) we have

$$\tau(^{137}_{55}\text{Cs}) \approx 2.433 \times 10^7 \, \text{s} \quad \approx 280 \, \text{days} \, .$$

4. Conclusion

Compton sources provide ultimately bright radiation for the nuclear forensics and the waste management. A specific property of Compton spectrum—steep high energy cutoff—enables to employ difference x-ray absorbtion imaging to detect the elements by its K-edge energy.

Application of the Compton radiation for transmutation of radioactive isotopes is not promising due to the small cross section of photo disintegration.

Acknowledgments

The work is supported by Photon and Quantum Basic Research Coordinated Development Program of MEXT: 'Fundamental Technology Development for High Brightness X-ray Source and the Imaging by Compact Accelerator.'

References

1. Bulyak E and Skomorokhov V 2004 *Proc. Eur. Part. Accel. Conf. 2004* report THPKF063
2. Bulyak E and Skomorokhov V 2005 *Phys. Rev. ST-AB* **8** 030703
3. Bulyak E and Urakawa J 2013 *Spectral properties of Compton inverse radiation: Application of Compton beams* http://arXiv/physics/1312.6785v1
4. http://www.nist.gov/pml/data/xraycoef/

Compact Intense Neutron Generators
Based on Inertial Electrostatic Confinement of D-D Fusion Plasmas[*]

K. Masuda[†], K. Inoue, T. Kajiwara and R. Nakamatsu

Institute of Advanced Energy, Kyoto University
Gokasho, Uji, Kyoto 611-0011, Japan
[†]*masuda@iae.kyoto-u.ac.jp*

A neutron generator based on inertial electrostatic confinement (IEC) of fusion plasmas is being developed for a non-destructive inspection system of special nuclear materials hidden in sea containers. The new IEC device is equipped with a multistage feedthrough which was designed aiming at both capability of a high bias voltage and enhancement of ion recirculation by modification of electric fields in the IEC device. Experimental comparison was made with a conventional single-stage IEC device developed in an earlier work. As the results, both the increase in the applied voltage and the modified field symmetry by the new multistage scheme showed significant enhancement in the neutron output. As a consequence, neutron output per input discharge current was enhanced drastically by a factor of ~30 in total. Also, the first pulsing experiments of the newly developed IEC neutron generator showed pulsed neutron output with a rapid pulse fall-off of ~ 1 μsec successfully.

Keywords: Compact neutron generator; inertial electrostatic confinement; nuclear fusion; active interrogation of special nuclear materials; counter terrorism.

1. Introduction

The concept of inertial electrostatic confinement (IEC) [1] was first proposed in 1950's aiming at the future fusion power plant. The first experiments [2] successfully demonstrated an ample neutron output of 10^9 D-T neutrons/sec in 1967 from an IEC driven by six ion guns. After a long pause of the research, the present new concept, IEC driven by a glow discharge (IEC-GD), came out in 1990's [3]. Since then, with its features of compactness and simplicity not requiring external ion sources, IEC-GD has been studied intensively for versatile applications as a neutron generatorby either D-T or D-D fusion reactions. In this report, two IEC-GD devices developed at Kyoto Universityare presented.

First, briefly introduced is a portable IEC-GD device which can produce more than 10^7 D-D neutrons/sec stably in dc operation with 80 kV and 80 mA.

---

[*] This work is supported by R&D Program for Implementation of Anti-Crime and Anti-Terrorism Technologies for a Safe and Secure Society promoted by Japan Science and Technology Agency.

This device was originally developed for landmine detection application [4], for which compactness was essential since weight of shielding material to be placed around the generator increases quadratically as the size of the generator increases. For this purpose, the IEC-GD deviceas compact as 25 cm in diameter was equipped with a water jacket for both cooling and shielding/moderating the energetic neutrons from fusion reactions.

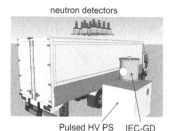

neutron detectors

Pulsed HV PS   IEC-GD

Fig. 1. A layout example for neutron-based SNMinterrogation.

Second, recent development of a pulsed IEC-GD neutron generator is described. Since 2010, we have developed an active non-destructive detection system for inspecting special nuclear materials (SNMs) such as ^{235}U in containers at sea ports [5, 6]. This consists of a primary neutron-based interrogation for a rapid screening purpose and a secondary gamma-ray-based stage for isotope identification vianuclear resonance fluorescence method using a quasi-monochromatic gamma-ray beam from laser Compton scattering. As the primary neutron-based system (see Fig. 1) is required to handle hundreds of sea containers per day, two advanced neutron-in neutron-out techniques are being developed, namely delayed neutron noise analysis (DNNA) and threshold energy neutron analysis (TENA) [6]. Pulsing of the IEC-GD neutron generator is mandatory for the former DNNA method. Another requirement for the neutron generator is as high averaged neutron yield as 10^8 neutrons/sec, which requires enhancement in the neutron yield by one order of magnitude compared with the aforementioned portable device. One of the most promising ways to enhance the neutron yield by an IEC-GD device is to apply a high bias voltage, because the fusion cross-section increases rapidly as the ion energy increases. In this context, the newly developed IEC-GD device is larger (~60 cm in diameter) than the aforementioned portable device (~35 cm in diameter), because portability is not mandatory in this case, while it is still transportable.

In general, a fusion-based neutron generator can run by either D-T or D-D;

$$D + T \rightarrow {}^4He + n\,(14.7MeV), \tag{1}$$

$$D + D \rightarrow {}^3He + n\,(2.45\,MeV). \tag{2}$$

The former has been and is being preferentially utilized for applications, because a higher neutron flux by more than two orders of magnitude can be obtained owing to its higher fusion cross-section than the latter. In contrast, we have chosen the use of D_2 rather than D_2/T_2 mixture as the fuel gas for both the landmine detection and SNM interrogation applications, because of advantages important for practical use, such as safety, easy maintenance without the need of tritium handling, and less shielding load for a lower incident neutron energy than D-T (see Eqs. (1) and (2)). In addition, the TENA method [6, 7] being developed for the SNM interrogation requires the less incident energy of neutrons from D-D reactions, in order to separate out the secondary fission neutrons from the background probing neutrons.

2. Compact DC IEC Neutron Generator

An IEC fusion device [1, 2] basically consists of a spherical anode at ground potential and a transparent gridded cathode at a negative potential of ~100 kV (see Fig. 2). Ions either provided by external ion sources or generated by a glow discharge (IEC-GD) [3] between the concentric electrodes are accelerated toward the center as they gain energy relevant to D-D, D-T and D-^{3}He fusion reactions from the applied electric fields.

The use of a transparent gridded cathode minimizes ion loss to the grid, and allows recirculation of energetic ions. Spherical focusing of ions is expected to result in a high ion density inside the gridded cathode. Those energetic ions thus collide with either ions or background neutrals, and undergo fusion reactions. The beam-beam fusion (that is, between two energetic ions) is most preferable as it is expected to show nonlinear increase of fusion rate as the ion current increases. However, the beam-beam fusion contribution is known to be negligibly small compared with the beam-gas fusion (that is, between an energetic ion and background neutral) in the current IEC devices. The main cause is rapid loss of ion energy by charge exchange collisions with background neutrals, particularly for IEC-GDs which require moderate operating gas pressure

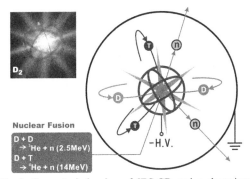

Fig. 2. Conceptual drawing of IEC-GD and a deuterium discharge photo.

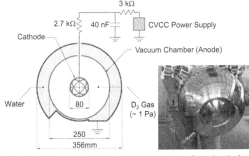

Fig. 3. An IEC-GD neutron generator for landmine detection application.

ranging from several tens mPa to ~ 1Pa.

Figure 3 shows schematic cross-section and a photo of a compact IEC-GD device of 350 mm in diameter (including 50 mm-thick water jacket), which we developed for landmine detection application. A water jacket has been designed for both cooling the anode and focusing the D-D neutrons toward the target, i.e. the minefield to enhance the incident thermal neutron flux into the mine [8].The water-cooling of the chamber enabled dc power of as high as 6 kW for a stable continuous operation for >8 hr. without falling off due to arcing and >1000 hr. total so far without maintenance. The neutron yield by this IEC-GD device is in excess of 10^7 D-D neutrons/sec (equivalent to more than 10^9 neutrons/sec when D-T gas is used) stably in dc operation with 80 kV and 80 mA[9].

An important advantage of IEC-GDs over neutron generators employing solid targets comes from the use of "gas target", which enables high-power and long-term dc operation.The IEC-GD also shows advantages of robustness, low cost, easy maintenance and easy operation owing to its extremely simple configuration, all of which are essential for practical applications.

3. Development of Intense Pulsed IEC Neutron Generator

3.1. *IEC Device Equipped with Multistage High-Voltage Feedthrough*

As is aforementioned, the current IEC-GD relies on beam-gas collision regime rather than beam-beam. In this regime, the fusion rate, dN/dt, is found to be expressed as follows,

$$\frac{dN}{dt} \propto P I \left\langle \int_0^L \sigma v dl \right\rangle, \qquad (3)$$

where P denotes D_2 gas pressure, i.e. the target density, and I denotes glow discharge current which corresponds to the flux of fast deuteriumions incident into the gas target. The bracket $<>$ indicates summation over the fast ions, and v and σ denotes the ion velocity and the fusion cross-section at that velocity, respectively. Finally, L denotes the pass length of ion, which is limited either by charge-exchange collision against background neutralsor loss to the feedthrough metal rod providing the bias voltage to the central cathode (see Fig. 3). The latter results from inherent asymmetry of the electrostatic fields induced by the existenceof the feedthrough at the high negative potential, which limits the envisagedion recirculation, i.e. oscillatory motion of ions within the anode (see Fig. 2), prior to striking the feedthrough rod.

A feature of IEC-GD tonote is that the operating gas pressure, P, and the discharge voltage, V, correlates with each other according to the Paschen's law, i.e. V is determined by a product Pd where d denotes the distance between the electrodes. As a consequence, increase of P results in decrease of V and accordingly both v and σ decrease in Eq. (3), resulting in a reduced fusion rate eventually. More importantly, use of a larger anode, i.e. increase in d, reduces P for a fixed V. Though this leads to a longer L in Eq. (3) depending almost linearly on the anode radius, the use of a larger anode does not result in increase of the fusion rate,

Fig. 4. An IEC-GD employing a 5-stage feedthrough.

because P reduces so that the product of P and the anode radius should be almost constant.

Although it does not change the neutron yield very much as such, we have chosen the use of a large anode of 560 mm in the new IEC-GD development, in order to ensure capability for a high voltage up to 200 kV.

Besides the use of the large anode, an important key technique incorporated in the new device is a multistage high-voltage feedthrough [10]. Figure 4 shows a photo of the device, consisting of a 560 mm spherical anode consisting of horizontal metal rings and a 200 mm spherical gridded cathode, both of which are to be placed in a cylindrical vacuum chamber of 600 mm inner diameter. A 5-stage feedthrough is used employing four intermediately biased cylindrical

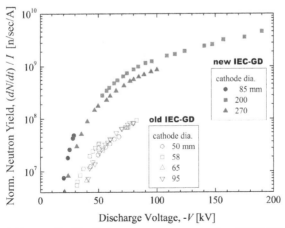

Fig. 5. Neutron yields normalized by discharge current, comparing the old IEC-GD (250 mm anode dia. with single-stage feedthrough) and the new one (560 mm anode dia. with 5-stage feedthrough) for various cathode diameters.

electrodes surrounding the innermost rod connected to the central cathode (more details of the geometry can be found in ref. 10).

The use of multistage feedthrough aims at capability for a high negative bias up to 200 kV. With a higher voltage, both v and, more significantly, σ in Eq. (3) are expected to increase as the ion energy increases. The present multistage feedthrough design [10] also aims to maximize the ion recirculation by modifying spherical symmetry of electric fields, which is also expected to enhance the neutron yield by increasing L in Eq. (3).

3.2. DC Experiments for Neutron Yield Comparison

Figure 5 shows experimental results with the newly developed IEC-GD employing the 5-stage feedthrough by use of a dc power supply, comparing neutron yield normalized by discharge current with the portable IEC-GD device presented in Sec. 2 (referred to 'old IEC-GD' in the figure). Note that the neutron yield is known to be proportional to the discharge current (see Eq. (3)). The new device shows capability for a high voltage up to 190 kV successfully. The new device is seen to produce more neutrons than the old one for the same voltage, roughly by a factor of 5–8. Again note that the use of the larger anode does not cause this enhancement as is discussed in Sec. 3.2, thus this enhancement would result from the envisaged improvement of the ion recirculation by the use of the new multistage feedthrough scheme. The increase in the voltage from 80 kV to 190 kV is seen to result in an additional enhancement in the neutron output by a factor of ~5. As a consequence, enhancement by a factor of ~30 was achieved, which eventually resulted in a high output in excess of 10^8 D-D neutrons/secfor a discharge current of 30 mA.

3.3. Preliminary Results from Pulsed Operation

Pulsed operation tests are being carried out by use of a 200 kV pulsed power supply show in Fig. 6. All circuit components in-cluding a dc charging power

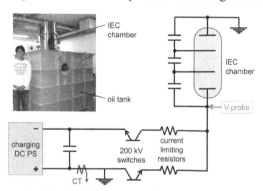

Fig. 6. Equivalent circuit of the IEC-GD employing 3-stage feedthrough and the 200 kV pulsed power supply developed for the SNM interrogation application.

supply, charging capacitor and two high-voltage switches shown in the equivalent circuit in Fig. 6 are setup in a single oil tank at ground potential to minimize emission of electro-magnetic noises which may reduce signal to noise ratio in detecting neutrons from SNMs. The IEC-GD device is equipped with a 3-stage feedthrough as is schematically shown in

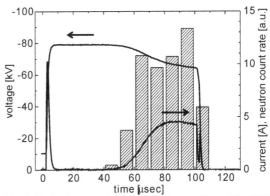

Fig. 7. Preliminary results from pulsed operation of the IEC-DG, showing temporal pulse shapes of applied charging voltage (80 kV), discharge current, and fast neutron count rate. The temporalresolution for the neutron count rate is 10 μsec.

the equivalent circuit. Because of the use of insulating oil for the pulsed system, we decided to reduce the number of intermediately biased electrodes for simplicity. A puled voltage up to 120 kV has been demonstrated so far, and further conditioning is being made for a higher voltage.

Preliminary results from pulsing experimentare shown in Fig. 7, showing temporal pulse shapes of applied voltage (80 kV), discharge current and neutron count rate by a fast neutron detector placed at ~3 m distance from the IEC-GD device. Though a considerable delay of tens μsec in the discharge current start-up (and accordingly in the fast neutron count rate) was seen and is to be studied, the pulse fall time wasfound as short as ~1μsecor less, and the discharge current pulse shape was seen very stable without significant shot-by-shot fluctuation.

4. Concluding Summary

IEC-GD neutron generators can produce an ample amount of neutrons from fusion reactions in an extremely simple and compact configuration. The compact IEC-GD of 25 cm in the anode diameter equipped with a conventional single-stage feedthrough produces more than 10^7 D-D neutrons/sec in dc mode. The new multistage feedthrough scheme resulted in a drastic enhancement in neutron output per current by a factor of ~30. The newly developed pulsed IEC-GD device for the SNM interrogation system showed promising features of a high neutron output in excess of 10^8 D-D neutron/sec in the dc tests, and stable pulsed output with a rapid pulse fall-off time of ~ 1μsec in the preliminary pulsed operation tests. Though these features were demonstrated with a

56 cmanode, the device size could be reduced in the future in order to extend its application, for example, toa portable SNM interrogation system.

References

1. P.T. Farnsworth, Electric Discharge Device for Producing Interactions between Nuclei, *U.S. Patent* 3258 402A (1966).
2. R.L. Hirsch, Inertial-Electrostatic Confinement of Ionized Fusion Gases, *J. Appl. Phys.* **38**, 4522 (1967).
3. J.H. Nadler et al., Characterization of an Inertial-Electrostatic Confinement Glow Discharge (IEC-GD) Neutron Generator, *Fusion Tech.* **21**, 1639 (1991).
4. H. Ohgaki et al., Conceptual Design of a Nuclear Material Detection System Based on the Neutron / Gamma-Ray Hybrid Approach, in *Proc. of 2010 IEEE International Conference on Technologies for Homeland Security (HST 2010)*, (Waltham, MA, USA, 2010) pp.525-529.
5. H. Ohgaki et al., Non-Destructive Inspection System for Special Nuclear Material using Inertial Electrostatic Confinement Fusion Neutrons and Laser Compton Scattering Gamma-rays, in *Proc. of 2012 IEEE International Conference on Technologies for Homeland Security (HST 2012)*, (Waltham, MA, USA, 2012) pp. 666-671.
6. K. Yoshikawa et al., Research and Development of the Humanitarian Landmine Detection System by a Compact Fusion Neutron Source, *IEEE Trans. Nuclear Sci.* **56**-3, 1193 (2009).
7. Y. Takahashi et al., Development of Active Neutron-based Interrogation System with D-D Neutron Source for Detection of Special Nuclear Materials, in these proceedings.
8. T. Takamatsu et al., Spatial Distribution of D-D Neutrons of a Compact Water-Cooled Inertial Electrostatic Confinement Device, *Fusion Sci. Tech.* **52**-4, 1114 (2007).
9. K. Masuda et al., Research and Development of Compact Neutron Sources based on Inertial Electrostatic Confinement Fusion, in *Proc. of 20th International Conference on the Application of Accelerators in Research and Industry (CAARI 2008)*, (Fort Worth, TX, USA, 2008) pp.587-590.
10. K. Masuda et al., Numerical Study of Ion Recirculation in an Improved Spherical Inertial Electrostatic Confinement Fusion Scheme by Use of a Multistage High-Voltage Feedthrough, *Fusion Sci. Tech.* **60**-2, 625 (2011).

NRF Based Nondestructive Inspection System for SNM by Using Laser-Compton-Backscattering Gamma-Rays[*]

H. Ohgaki[†], M. Omer, H. Negm, I. Daito, H. Zen, T. Kii, K. Masuda and T. Hori

Institute of Advanced Energy, Kyoto University
Gokasho, Uji, Kyoto 6110011, Japan
[†]*ohgaki.hideaki.2w@kyoto-u.ac.jp*

R. Hajima, T. Hayakawa, T. Shizuma and M. Kando
Quantum Beam Science Directorate, Japan Atomic Energy Agency, Tokai, Ibaraki, 319-1195, Japan.

A non-destructive inspection system for special nuclear materials (SNMs) hidden in a sea cargo has been developed. The system consists of a fast screening system using neutron generated by inertial electrostatic confinement (IEC) device and an isotope identification system using nuclear resonance fluorescence (NRF) measurements with laser Compton backscattering (LCS) gamma-rays has been developed. The neutron flux of 10^8 n/sec has been achieved by the IEC in static mode. We have developed a modified neutron reactor noise analysis method to detect fission neutron in a short time. The LCS gamma-rays has been generated by using a small racetrack microtoron accelerator and an intense sub-nano second laser colliding head-on to the electron beam. The gamma-ray flux has been achieved more than 10^5 photons/s. The NRF gamma-rays will be measured using $LaBr_3(Ce)$ scintillation detector array whose performance has been measured by NRF experiment of U-235 in HIGS facility. The whole inspection system has been designed to satisfy a demand from the sea port.

Keywords: Non-destructive inspection system; special nuclear material; IEC neutron source; neutron reactor noise analysis method; LCS gamma-rays; $LaBr_3(Ce)$ detector.

1. Introduction

An active non-destructive detection system for special nuclear materials (SNMs) such as ^{235}U has been developed for the container inspection at the sea ports in Japan [1, 2]. In general, the interrogation technique which uses neutrons as the

[*] This work is supported by R&D Program for Implementation of Anti-Crime and Anti-Terrorism Technologies for a Safe and Secure Societypromoted by Japan Science and Technology Agency.

probe has a high sensibility for hidden SNMs and promises a short inspection time. However, it is difficult to separate the probe neutrons and the fission based neutrons. Thus the positive failure in the detection of SNMs could be large. On the other hand, gamma-ray based interrogation system which employs Nuclear Resonance Fluorescence (NRF) technique has a potential of excellent isotope determination ability. However, because the NRF yields from SNMs are scarce, a long inspection time could be required for a gamma-ray based system. Therefore, our developing inspection system consists of a fast screening system by using Inertial Electrostatic Confinement (IEC) neutron source [3, 4] and an isotope identification system by using NRF [5] with the Laser Compton (LCS) gamma-rays [6-8]. In this report, we briefly introduceour SNM inspection system.

2. Fast Screening System

The required neutron generators for an inspection system of hidden SNMs in a sea cargo should give a high neutron flux of 10^8 neutrons/sec or more with robust safety. There are two types of neutron generation reactions, D-T and D-D. The neutron flux from the D-T reaction is 100 times higher than that from the D-D reaction. However, for the safety reason we chose the D-D reaction, because the tritium isa radioactive material. In this line, we need to increase the neutron flux from the D-D source. A possible solution could be an IEC of fusion plasma. Because an advantage of the IEC over beam-target type neutron generator is capability of high-power operation owing to the use of gas-target. For instance, 6.4 kW dc discharge power to produce D-D neutrons in excess of 10^7 neutrons/sec [9] and applying a higher bias voltage could enhance the neutron yield. The multistage high-voltage feedthrough scheme [10] has been developed and we achieved the neutron flux of 10^8 neutrons/sec by applying 190 kV as shown in Figure 1.

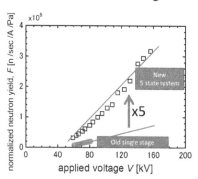

Fig. 1. Neutron yield normalized by operating gas pressure and discharge current as a function of bias voltage.

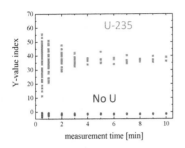

Fig. 2 Neutron yield normalized by operating gas pressure and discharge current as a function of bias voltage.

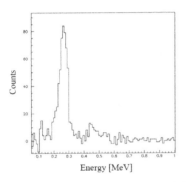

Fig. 3 Measured gamma-rays generated by laser Compton scattering.

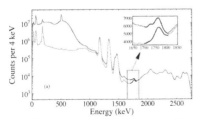

Fig. 4 NRF spectrum and the background (off-beam) spectrum measured by the LaBr₃(Ce) detector array.

To detect the fission base neutron, we have developed the modified neutron reactor noise analysis method [11] which focuses on the delayed neutron components to decrease the effect of the fluctuation of neutron source. Figure 2 shows the result of the experiments by using HEU plates of 0.5 kg irradiated by pulse neutron of 2×10^5 n/s. We used 4 He-3 neutron detectors surrounded the fuel assemblies. The red points indicate the Y-value, which indicates a neutron counting rate variation from the Poisson distribution, in case of existence of U-235 in the core. The result clearly shows that we can detect U-235 even in a short measurement time less than 3 minutes. By using this experimental result, we design the measurement system for the fast screening system.

3. Isotope Identification System

Several novel inspection methods have been proposed and examined using gamma-rays. Bertozzi et al. have proposed the NRF measurement by using bremsstrahlung radiations [5]. The NRF gamma-rays from irradiated nucleus can be used for an isotope identification, because the energy of the NRF gamma-rays uniquely determined as a finger print of

206

specific nuclear species. We have proposed to use a monochromatic gamma-ray LCS process to obtain good S/N ratio. To make a system compact, our developing system employsthe LCS gamma-ray source from a 240-MeV microtron accelerator [12]. We have demonstrated generation of a LCS gamma-ray beam using a 150-MeV microtron electron accelerator at Kizugawa, JAEA [13]. The LCS gamma-ray beam was generated by a nearly head-on collision between the 150 MeV electrons and laser photons provided from a Nd:YAG system whose wavelength is 1064 nm and pulse energy of 300 mJ, 200 ps pulse width compressed by a stimulated Brillouin scattering cells [14]. The total flux of the generated LCS gamma-rays was confirmed by Compton scattering photons from an Al target with a thickness of 3 mm or 0.5 mm. The scattered photons were measured using a GSO scintillator with a size of 20 mm x 20 mm x 50 mm (Figure 3). The evaluated LCS gamma-ray flux isabout 10^5 photons/s when we use a laser pulse of 1 J with a frequency of 10 Hz.

For the NRF gamma-ray measurement system, we have performed an experiment to examine the feasibility of the NRF detection system by using $LaBr_3(Ce)$ in combination with the LCS gamma-rays at Duke University [15]. A circularly polarized gamma-ray beam with an energy of 1733 keV irradiated to a U-235 (with 93.7 % enrichment) target whose total mass was 4.62 g. The target was mounted in an evacuated pipe to decrease the background that may result from the air scattering. Two cylindrical collimators of 15.2 cm and 25.4 cm long with 1.905 cm diameter collimated the incident gamma-ray beam to define the energy spread of approximately 5% (FWHM). Anarray detector consisted of eight $LaBr_3(Ce)$ detectors of 1.5" in diameter and 3" in length. Figure 4 shows the NRF (on-beam) spectrum and the background (off-beam) spectrum. The NRF peak at 1733 keV is clearly observed[16].

4. Active Non-Destructive Detection System

The developing detection system consists of fast screening system by using IEC neutrons and isotope identification system by using NRF-laser Compton gamma-rays. During the study, we found that the fast screening system would not specify the

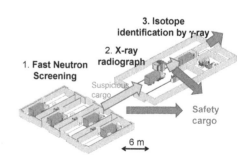

Fig.5.Non-destructive SNM inspection system.

position of suspicious material, when we set the inspection period to be a few minutes. Therefore we plan to install an X-ray transmission radiograph system to specify the suspicious position in the cargo where the LCS gamma-ray will irradiate to identify the hidden SNMs. Figure 5 illustrates the developing inspection system. The inspection system is designed to detect 1 kg HEU hidden in a 20 TEU in 2 minutes by the fast screening system with 10% positive failure and 10 minutes by the isotope identification system.

References

1. H. Ohgaki *et al.*, "Conceptual design of a nuclear material detection system based on the neutron / gamma-ray hybrid approach," *Proc. of IEEE HST* 2010, pp. 525-529, (2010).
2. H. Ohgaki *et al.*, "Non-Destructive Inspection System for Special Nuclear Material using Inertial Electrostatic Confinement Fusion Neutros and Laser Compton Scattering Gamma-rays," *Proc. of IEEE HST* 2012, pp. 666–671, (2012).
3. R.L. Hirsch, "Inertial-Electrostatic Confinement of Ionized Fusion Gases,"*J. Appl. Phys.* 38, 4522, (1967).
4. G.H. Miley, L. Wu and H.J. Kim "IEC-based neutron generator for security inspection system," *J. Radioanalytical Nucl. Chem.* 263, 159, (2005).
5. W. Bertozzi and R. Ledoux, Nucl. Instru. Metho. B, 241, 820, (2005).
6. J. Pruet, D. P. McNabb, C. A. Hagmann, F. V. Hartemann, and C. P. J. Barty, *J. Appl. Phys.* 99, 123102, (2006).
7. N. Kikuzawa, R. Hajima, N. Nishimori, E. Minehara, T. Hayakawa, T. Shizuma, H. Toyokawa, and H. Ohgaki, *Appl. Phys. Express* 2, 036502, (2009).
8. T. Shizuma, T. Hayakawa, R. Hajima, N. Kikuzawa, H. Ohgaki, and H. Toyokawa, *Rev. Sci. Instru.*, 83, 015103, (2012).
9. K. Yoshikawa et al., IEEE Trans. Nucl. Sci. 56, 1193, (2009).
10. K. Masuda et al., Fusion Sci. Technol. 60, 625, (2011).
11. T. Misawa, Y. Takahashi, Y. Yamaguchi, T. Yagi, C. H. Pyeon, K. Masuda, T. Kajiwara, H. Ohgaki, Proc. of 2012 IEEE Nuclear Science Symposium and Medical Imaging Conference, Anaheim, California, USA, Oct. 29–Nov. 3, 2012, CD-ROM (2012).
12. T. Hori, T. Kii, R. Kinjo, H. Ohgaki, M. Omer, H. Zen, I. Daito, R. Hajima, T. Hayakawa, M. Kando, H. Kotaki, Proc. of the IPAC2012, 4127 (2012).
13. K. Kawase et al., Rev. Sci. Instrum. 79, 053302 (2008).

14. I. Daito, R. Hajima, T. Hayakawa, Y. Hayashi, M. Kando, H. Kotaki, T. Shizuma, H. Ohgaki, Proc. IPAC-2012, 4124 (2012).
15. V. N. Litvinenko et al., Phys. Rev. Lett. 78, 4569, (1997).
16. M. Omer, H. Negm, H. Zen, I. Daito, T. Kii, K. Masuda, H. Ohgaki, R. Hajima, T. Shizuma, T. Hayakawa, N. Kikuzawa, Japan Journal of Applied Physics, vol. 52, 10, 106401-1-4 (2013).

Development of Measurement Methods for Detection of Special Nuclear Materials using D-D Pulsed Neutron Source

Tsuyoshi Misawa*, Yoshiyuki Takahashi, Takahiro Yagi and Cheol Ho Pyeon

Research Reactor Institute, Kyoto University
Kumatori, Osaka 590-0494, Japan
*misa@rri.kyoto-u.ac.jp

Masaharu Kimura

Graduate School of Energy Science, Kyoto University
Yoshida, Kyoto 606-8501, Japan

Kai Masuda[†] and Hideaki Ohgaki

Institute of Advanced Energy, Kyoto University
Uji, Kyoto 611-0011, Japan
†masuda@iae.kyoto-u.ac.jp

For detection of hidden special nuclear materials (SNMs), we have developed an active neutron-based interrogation system combined with a D-D fusion pulsed neutron source and a neutron detection system. In the detection scheme, we have adopted new measurement techniques simultaneously; neutron noise analysis and neutron energy spectrum analysis. The validity of neutron noise analysis method has been experimentally studied in the Kyoto University Critical Assembly (KUCA), and was applied to a cargo container inspection system by simulation.

Keywords: Special nuclear materials; pulsed neutron; neutron noise analysis; DNNA.

1. Introduction

The detection of hidden special nuclear materials (SNMs) is an important issue for nuclear security to counter terrorist threats The interrogation system to detect SNMs used in a seaport or an airport are under developing by many researchers. In active interrogation for SNMs by neutron injection into a target, it can be accomplished by the use of either D-D or D-T neutron source to generate pulses of neutrons. Neutrons produced by those neutron sourcesare thermalized and serve to initiate fission events in SNMs. In the neutron and/or gamma analysis methodologies, the presence or absence of SNMs can be detected. One of conventional neutron-in neutron-out detection techniques is called Delayed

Neutron Analysis (DNA) [1], which is based on the detection of delayed neutrons from precursor nuclides produced by induced fission reactions. The other conventional neutron-based technique called Differential Die-Away Analysis (DDA) [2] makes use of the premise that, when a fissile material is exist, fast neutron population outside of a cargo decays with a decay time of surrounding media, i.e., the thermal die-away time in the cargo. Both the techniques however fail to make use of the major part of prompt neutrons emitted during the incident neutron pulses, and therefore we require highly intense pulses of interrogating neutrons. A D-T neutron generator would be thus mandatory for the cargo screening through those techniques, because it provides the neutron flux two orders of magnitude higher than those from D-D reactions, though the use of the radioactive tritium is not favorable in the viewpoint of safety, easy operation, and maintenance of the inspection system.

In this study, we propose a neutron-detection scheme based on a neutron noise analysis technique which is called Delayed Neutron Noise Analysis (DNNA), in conjunction with the use of a D-D neutron source based on the inertial electrostatic confinement (IEC) of fusion plasmas [3]. Note that the detail explanation about this D-D neutron source [4] is described in the other paper in the proceeding of this meeting presented by K. Masuda.

2. Delayed Neutron Noise Analysis (DNNA) Method

The neutron noise analysis methods were developed originally in the reactor physics field. The variance-to-mean value method, called Feynman-alpha method, is one of well-known noise analysis methods to measure subcriticality of reactors [5]. In this method, the variation of neutron count rate distribution from the Poisson distribution is measured because if there happen the fission chain reactions by fissile materials, the neutron count rate distribution does not show the Poisson distribution even by using an isotope neutron source such as an Am-Be, which means the existence of fissile materials in the target system. Figure 1 shows a typical example of measured data by the Feynman-alpha method in a subcritical reactor using a continuous external neutron source; an Am-Be isotope neutron source, where the horizontal axis shows the gate width of measured time and the vertical axis shows the Y-value that is an index to express the degree of the variation of the neutron count rate distribution from the Poisson distribution. If the Y-values are greater than zero, it indicates the existence of fission chain reactions by fissile materials. The theoretical formula for the Y-value based on one point reactor approximation is inserted in Fig. 1, where alpha is the prompt neutron decay constant in the target subcritical system.

In a near critical state, it is easy to determine the existence of chain fission reactions by the Feynman-alpha method. However, it is difficult to measure the Y-value in deep subcritical state such as a system for detection of SNMs, because number of emitted by fission chain reaction is small which causes large statistical error and the results are largely affected by the location of the detector. Moreover, the neutron count rate is very high after injection of pulsed neutron and it decreases quickly in the subcritical state as shown in Fig. 2. In this situation, the dead time effect of neutron detectors in the prompt region cannot be ignored. To overcome these problems, a new noise analysis method called Delayed Neutron Noise Analysis (DNNA) was developed. In this new method, the neutron counts measured only in the delayed neutron region after injection of pulsed neutrons (see Fig. 2) were analyzed to obtain the Y-values in the Feynman-alpha method, whereas the prompt region was completely ignored.

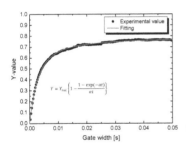

Fig. 1 Experimental results of Feynman-alpha method.

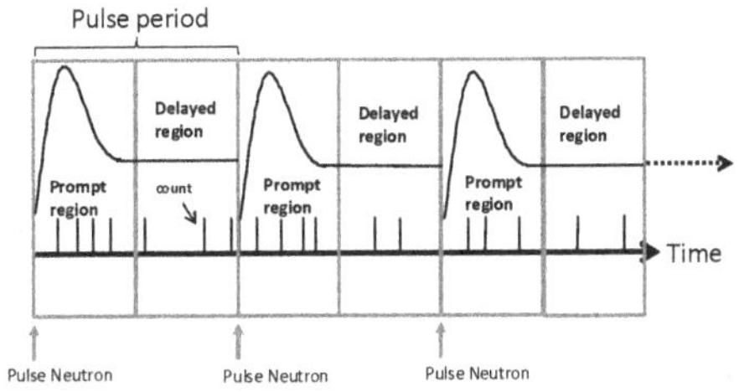

Fig. 2 Explanation of prompt region and delayed region after injection of pulsed neutron.

3. Experiments at KUCA

The validity of the present new DNNA method was experimentally investigated in the Kyoto University Critical Assembly (KUCA) at Kyoto University Research Reactor Institute. KUCA is the only one system where SNMs detection experiments can be carried out using enriched uranium fuel in Japan. The system shown in Fig. 3 is composed of several fuel assemblies that contain highly enriched uranium or natural uranium fuel plates. Pulsed neutrons with an energy of 14 MeV by the D-T reactions are injected into the fuel assemblies. The neutrons are provided from a Cockcroft-Walton type accelerator installed in the KUCA as shown in Fig. 4 [6]. Neutron counts are measured by several ³He neutron detectors placed apart from the fuel region and detection data signals are acquired by a time train data acquisition system with a time resolution of 100 nsec.

Fig. 3 Experimental configuration for SNM detection. Yellow line corresponds to the fuel region containing highly enriched uranium.

Fig. 4 D-T neutron production accelerator at KUCA.

Figure 5 shows the experimental results to measure the Y-values with and without highly enriched uranium. Note that in the present experiment, the weight of highly enriched uranium is approximately 0.5 kg. Nine 1-inch-diameter ³He neutron detectors are used. The neutron production intensity by the D-T reactions is about 2×10^5 n/sec. The measurement time is 30 min. The neutron pulse period and the pulse width are 100 msec and 100 micro sec, respectively. It was found that the Y-values were almost zero in the case of without uranium fuel. However, these values are clearly larger than zero in the case of existence of uranium fuel assemblies. The Y-value used as an index of existence of fissile chain reactions becomes larger with an increase of number of neutron detectors adopted in the neutron noise analysis, which means increase of detection efficiency. In this experiment, the D-T neutron source was adopted because of administrative reason. However, it is easy to apply this method with use of a D-D neutron source, and existence of SNMs could be determined by the present neutron noise analysis method with use of delayed neutrons even by using the D-D neutron source.

As mentioned in the previous section, if the Y-values obtained in the DNNA are greater than zero, we can judge that fissile materials, namely SNMs, exist in the target system such as a cargo container. However, when the measurement time is short, the total number of neutron counts is small and the Y-values show a large fluctuation because of a statistical reason and sometimes become less than zero even in case of existence of SNMs. To determine the minimum required measurement time for detection of SNMs, we define as follows; when all Y-values that are the function of gate width of measured time become greater than zero, we can conclude that fission chain reactions, namely existence of SNMs, are detected. According to this definition, in case of the experimental results shown in Fig. 5, the existence of fissile materials can be detected by 3 minutes measurement time in the DNNA method with use of one ^{3}He neutron detector.

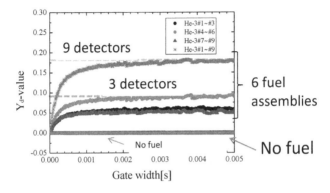

Fig. 5 Measured results by the present DNNA method with highly enriched uranium fuel by changing number of neutron detectors in noise analysis. The results without fuel assembly are also presented.

4. Simulation for Cargo Container Inspection

For a practical case, the present DNNA method was applied to cargo container inspection system for SNMs detectionand simulation was carried out by MCNP Monte Carlo calculation. Figure 6 shows the arrangement of the system viewed from the back front of the 20-foot container, which consists of anIEC D-D neutron source, stack of neutron detectors located above the cargo container,and neutron reflector walls made by polyethylene surrounding the container.

We estimated the number of 1-inch-diameter ^{3}He neutron detectors required to detect the SNM at the center of the container as shown in Fig. 6 with the 10-minute measurement time according to the definition mentioned in the previous section. Note that neutron production rate from the D-D pulsed neutron source is assumed to be 1×10^8 n/sec.

The result of the simulation is shown in Fig. 7. It was found that for detection of highly enriched uranium of weights of 1kg and 5kg, the minimum required number of 1-inch-diameter ^{3}Heneutron detectors is about 32 and 12, respectively. We can conclude that this number of the detectors is feasible in a real inspection system.

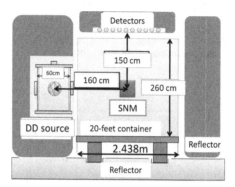

Fig. 6 Arrangement of the cargo container inspection system.

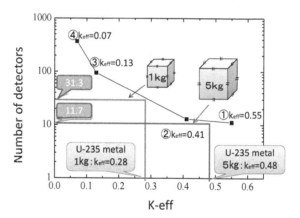

Fig. 7 Required number of neutron detectors for SNM detection. The horizontal axis, K-eff, indicates neutron multiplication a factor of the system. The vertical axis shows the required number of detectors.

5. Conclusion

We have developed an active neutron-based interrogation system combined with an IEC D-D fusion pulsed neutron source and neutron detection system for detection of SNMs hidden in a cargo container. For this purpose, we developed a new method; modified neutron noise analysis method based on the Feynman-alpha method applied to the delayed neutron region which is appeared after injection of pulsed neutrons; called DNNA method. Experiments at the KUCA reactor and calculations by Monte Carlo simulations are carried out. These results show that this system has an ability to detect SNMs and it can be applied to real cargo container inspection systems. Further experiments and analysis will be carried out to optimize the system arrangements; a D-D neutron source, detectors, reflectors and so on.

Acknowledgments

This work is partially supported by Japan Science and Technology Agency and JSPS KAKENHI Grant Number 25420906.

References

1. R. F. Radel et al., Fusion Sci. and Technol. 52 (2007) 1087.
2. K. A. Jordan et al., Nucl. Intrum. Meth. A 579 (2007) 388.
3. G.H. Miley et al., J. Radioanalytical Nucl. Chem. 263 (2005) 159.
4. K. Yoshikawa et al., IEEE Trans. Nucl. Sci. 56 (2009) 1193.
5. T. Misawa et al., Nucl. Sci. Eng., 104 (1990) 53.
6. C. H. Pyeon et al., Ann. Nucl. Energy, 40 (2012) 229.

SOFIA, a Next-Generation Facility for Fission Yields Measurements and Fission Study. First Results and Perspectives

L. Audouin[1,2] *, E. Pellereau[3,1], J. Taieb[3], G. Boutoux[3], G. Béliera[3], A. Chatillon[3], A. Ebran[3], T. Gorbinet[3], B. Laurent[3], J.-F. Martin[3], L. Tassan-Got[2,1], B. Jurado[8], H. Alvarez-Pol[4], Y. Ayyad[4], J. Benlliure[4], M. Caamano[4], D. Cortina-Gil[4], B. Fernandez-Dominguez[4], C. Paradela[4], J.-L Rodriguez-Sanchez[4], J. Vargas[4], E. Casarejos[5], A. Heinz[6], A. Kelic-Heil[7], N. Kurz[7], C. Nociforo[7], S. Pietri[7], A. Prochazka[7], D. Rossi[7], K.-H. Schmidt[7], H. Simon[7], B. Voss[7], H. Weick[7] and J.S. Winfield[7]

[1] *Université Paris-Sud, 91405 Orsay cedex, France*

[2] *CNRS, IPN Orsay, F-91406 Orsay, France*

[3] *CEA, DAM, DIF, F-91297 Arpajon, France*

[4] *Universidade de Santiago de Compostela, E-15782 Santiago de Compostela, Spain*

[5] *Universidade de Vigo, E-36310 Vigo, Spain*

[6] *Chalmers University of Technology, SE-412 96 Gothenburg, Sweden*

[7] *GSI, Gesellschaft fur Schwerionenforschung GmbH, Planckstr. 1, D-64291 Darmstadt, Germany*

[8] *CNRS, CENBG, F-33175 Gradignan, France*

Fission fragments play an important role in nuclear reactors evolution and safety. However, fragments yields are poorly known : data are essentially limited to mass yields from thermal neutron-induced fissions on a very few nuclei. SOFIA (Study On FIssion with Aladin) is an innovative experimental program on nuclear fission carried out at the GSI facility, which aims at providing isotopic yields on a broad range of fissioning systems. Relativistic secondary beams of actinides and pre-actinides are selected by the Fragment Separator (FRS) and their fission is triggered by electromagnetic interaction. The resulting excitation energy is comparable to the result of an interaction with a low-energy neutron, thus leading to useful data for reactor simulations. For the first time ever, both fission fragments are completely identified in charge and

---

*Corresponding author : audouin@ipno.in2p3.fr, http://ipnweb.in2p3.fr

mass in a new recoil spectrometer, allowing for precise yields measurements. The yield of prompt neutrons can then be deduced, and the fission mechanism can be ascribed, providing new constraints for fission models. During the first experiment, all the technical challenges were matched : we have thus set new experimental standards in the measurements of relativistic heavy ions (time of flight, position, energy loss).This communication presents a first series of results obtained on the fission of ^{238}U; many other fissioning systems have also been measured and are being analyzed presently. A second SOFIA experiment is planned in September 2014, and will be focused on the measurement of the fission of ^{236}U, the analog of ^{235}U+n.

Keywords: Nuclear fission; fission fragments; nuclear data; isotopic yields; nuclear energy.

1. Context and aims

A precise description of nuclear fission is of major interest to improve the safety of nuclear reactors and to assess the potentialities of new types of reactors and/or fuels. Fission fragments play a key role in reactors: they are the vector of most of the energy liberated during fission; they are the source of the delayed neutrons, which the stability of the chain reaction largely depends on; and they are responsible for the poisoning of the core, a point which may become even more important in the future if high burn-up fuels are used as the poisoning increases practically linearly with the burn-up. Indeed, increasing burn-up is a constant trend of the nuclear power industry: while the first fuels were designed for burn-ups around 30 MWj/T, values of 60 are now getting common, and numbers like 80 or 90 are often mentioned as future goals. The role of fission fragments is also key in the case of a reactivity excursion, as their inventory suddenly increases with the core power. Finally, fission fragments play a central role in the residual thermal power of cores, which is acknowledged as the major source of operational risk, a point tragically emphasized by the Fukushima incident.

Beyond its applicative aspects, fission is a complex process which highlights many nuclear properties. Despite being one of the longest-studied nuclear reactions, no theoretical approach is yet able to describe the whole fission process, i.e. properties of the fissioning system, fission dynamics and fission fragment distributions. Further advances in our understanding of the fission mechanism depend of course on theoretical developments, but also on our ability to access to new observables, especially the isotopic distributions of both fission fragments in coincidence or a very precise measurement of their kinetic energy.

When it comes to fission fragments, data are surprisingly scarce. As already pointed out, there is no precise data for the isotopic yields of fission fragments. Even for the mass yield of a supposedly well-know reaction (^{235}U+n$_{th}$), experimental uncertainties remain in the order of 15% for most isobars, and the evaluations tend to exhibit significant differences. Only a handful of isotopes have their yield precisely measured (by gamma-spectroscopy). Up to recently, this lack of information was *de facto* considered as acceptable for the energy applications, because the calculation power for simulation codes did not allow to precisely handle the production of the hundreds of possible fragments. Pseudo-fragments and integral ingredients were used instead to approximate the core evolution. However, with the stronger emphasize on safety, the evolution toward larger burnups, and the growing CPU power, more refined calculations can now be undertaken, and new data are eagerly expected.

The lack of data about fission fragments is largely caused by the intrinsic difficulty of fission fragment measurements. In standard fission experiments, fissions are triggered by a projectile (often a neutron) in an actinide target - we'll call this "direct kinematics". In such experiments, only the masses of both fragments or the charge of the light one have ever been measured. The fission fragments energy is so small that it is hardly conceivable to get more information in this type of experiment. An answer to this limitation is the "inverse kinematics" method, in which an actinide beam interacts with a target to form excited systems which fission in-flight. As both fission fragments fly with the additional energy of the center of mass, a complete identification of fragments in a recoil spectrometer becomes possible. Furthermore, the inverse kinematics creates the opportunity to study very exotic actinides and pre-actinides, for which no fixed target could even be thought of. This technique has been successfully used at GSI since the end of the 90's[1] and more recently at GANIL[2]. But up to now, only one of the fragments was fully identified, or only the charge of both was determined. The simultaneous identification of both fragments had simply never been performed.

The SOFIA (Studies On FIssion with Aladin) collaboration was started in 2008 in order to realize a new generation of experiment on fission at GSI, taking advantage of the intense ^{238}U beam available at this facility with an energy up to 1 GeV per nucleon. The aim of the SOFIA experiments is to fill several of the aforementioned gaps in experimental data: to realize a complete isotopic identification of both fission fragments, to measure isotopic fission yields and fragments kinematics with a high precision, and to measure the neutrons yields in coincidence with fragments. A striking feature is that these data should be obtained not only on a few very long-lived isotopes (^{235}U, ^{238}U, etc.), but on a large variety of actinides and pre-actinides, even very exotic ones. In order to fulfill these objectives, we designed and built several detectors, thus creating a new high-resolution, large-acceptance recoil spectrometer around the pre-existing ALADIN magnet. This set-up can be - and has indeed also been - used to study collision-induced fission (for example proton-induced fissions, using a liquid hydrogen target[3]), resulting in excitation energies of tens and even hundreds of MeV. But the main design of SOFIA is to use an electromagnetic interaction to trigger fission by exciting the Giant Dipolar Resonance (GDR). In this case, the excitation energy is in most cases between 10 and 15 MeV, a range reasonably close to the application domain and for which extrapolation to the thermal neutron-induced fission is definitely possible.

2. Experimental system

One of the strengths of the SOFIA set-up is its ability to use the GSI fragment separator, the FRS[4], as a source of secondary beam. Practically, we sent the 1 AGeV ^{238}U beam delivered by the SIS synchrotron to a Be target at the entrance of the FRS. Some nuclei produced by fragmentation were selected by 3 of the 4 magnetic sections of the FRS, then sent in the cave C where the SOFIA set-up was installed.

The fissions of the actinides were triggered by an active target, which consisted in several layers of high-Z material (2 uranium foils and 1 lead foil). The intervals between layers were used as miniature ionization chambers. The combined energy loss of both fragments is roughly half the beam's one: hence, the large ΔE change caused by the fission offers an unambiguous information on the target layer in which the fission took place. Such an active target allows to increase the fission rate without sacrificing the resolution on the velocity of the fragments, as the necessary corrections can be accurately calculated. This is mandatory for a precise measurement of the total kinetic energy released in the fission.

In order to identify fission fragments as well as the nuclei which fissioned to form them, we used the same method : the nuclear charge was deduced from the energy loss in ionization chambers, and the mass was then obtained through the well-known magnetic deviation formula :

$$B\rho \propto \frac{A}{q}(\beta\gamma)$$

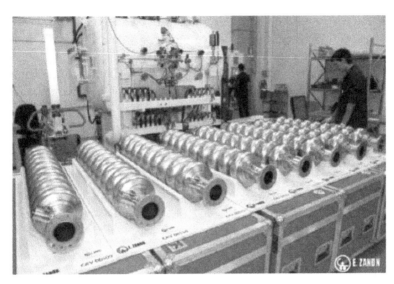

Fig. 1. Scheme of the SOFIA detection set-up for fission fragments. The detectors dedicated to the secondary beam identification, located upstream of the active target, have been left out for clarity.

where B is a magnetic field; ρ is the radius of the trajectory in the magnetic section, deduced from position measurements; q is the effective charge of the ion (equal to the nuclear charge most of the time due to their very large velocity); and β and γ are the relativistic Lorentz factors, deduced from a time-of-flight measurement. Figure 1 represents a scheme of the SOFIA set-up. The detectors dedicated to the secondary beam identification (2 ionization chambers, 2 Time Projection Chambers and 2 plastics) were located just upstream of the target, with an additional plastic in the middle of the FRS ; the magnetic element consisted in the last parts of the FRS. In the case of the fission fragments identification, we used the large acceptance ALADIN magnet. We designed and used a special ionization chamber, divided in two parts in order to accommodate the multiplicity of the fission, to obtain the Z. This chamber was also able to provide a very precise measurement of the horizontal position and angle of both fragments. Positions were also measured upstream and downstream from the magnet by 2 Multi-Wire Proportional Counters. Plastics detectors provided time-of-flight measurement.

With several of these detectors, the SOFIA collaboration has defined new performance standards for heavy ions physics. The times of flight were measured with a resolution of 40 ps (full width half maximum)[5], horizontal positions were determined with a resolution of 50 μm (FWHM), and the resolving power on energy loss was 1.2%. Thanks to these state-of-the-art performances, we fully resolved the nuclear charge of all fission fragments, and measured their masses with a resolution ranging between 0,6 and 0,8 nuclear mass, allowing for a clean extraction of isotopic yields. The raw spectrum in charge and mass for the fission of ^{238}U are presented in figure 2.

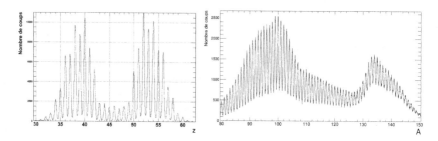

Fig. 2.   Raw charge (left) and mass (right) spectrum for the fission of ^{238}U.

Note that in this experiment, not all fissions are caused by electromagnetic interactions: some are due to nuclear collisions. The contribution of these fissions is suppressed in two steps. First, all the events for which the sum of the nuclear charges of the fragments is not equal to the charge of the secondary beam are rejected. The yields are then corrected using an appropriate scaling of the yields obtained in aluminium foils, where only nuclear collisions trigger fissions.

3. Experimental results

Unless otherwise stated, the results presented in this section come from the fission of ^{238}U. Data on this nuclei were mostly analyzed by Eric Pellereau[6].

The elemental and isotonic yields, plotted in Fig. 3, present the double-hump structure associated with low-energy fission. This has been interpreted for a very long time as a signature of shell effects, namely the closed neutron shell N=82 which acts as an attractor during the mass splitting. With the conservation of the N/Z ratio, this leads to the formation of a fragment close to the mass 134 regardless of the fissioning system, the other fragment taking over the rest of the nucleons. This influence of the shell is clearly visible on the isotonic spectra, with a production increased by

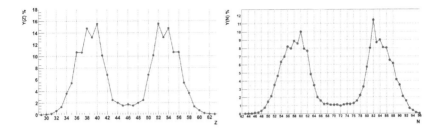

Fig. 3. Elemental (left) and isotonic (right) yields for the fission of ^{238}U.

about 25% with respect to neutron numbers 81 or 83.

For both protons and neutrons, even-odd effects are clearly visible. It can be quantified to 10% for Z and 5% for N. It is however interesting to compare the evolution of these even-odd effects as a function of the excitation energy of the prefragment. Due to the GDR excitation used in SOFIA, the average excitation energy of the fissioning system is 13 MeV, while a thermal neutron brings about 6 MeV. The Lohengrin facility has produced some thermal neutron data about even-odd effect on the fission of ^{239}U. In this case, the amplitude of the even-odd effect was found to be 5% for neutrons, and 24% for protons. Thus the even-odd effect is independent of the energy for the neutrons, while it strongly depends on it for the protons.

The mass yield is a well-known illustration of fission studies. In Fig. 4, we compare the yields obtained by SOFIA to the ENDF evaluation for 14 MeV neutrons (hence the 20 MeV fission of ^{239}U). The minor differences of the yields can clearly be ascribed to the different fissioning systems : the symmetric fission is slightly more likely at 20 MeV than at 13 MeV, and the additional neutron leads to a 1 mass unit displacement of the light fragment mass, while the heavy mass bump remains unchanged, as expected from the shell effect. Note that the error bars are present in both plots: but the SOFIA error bars are limited to 2-3%, while the ENDF evaluation has uncertainties around 15%. SOFIA brings data of unprecedented precision and therefore much stronger constraints for the fission modeling.

Figure 5 presents the yields as a function of the mass for each element of the heavy bump. Such data have never been obtained for low excitation energies. Again, error bars are so small they can barely be seen. A very interesting point is the radical change of shape between In, Sn and Sb distributions. Actually, these distributions can be interpreted as the sum of

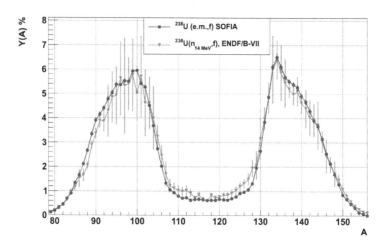

Fig. 4.   Mass yields for the electromagnetic-induced fission of ^{238}U, compared to the 14 MeV neutron-fission of ^{238}U.

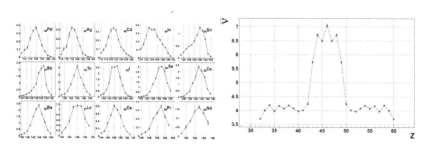

Fig. 5.   Isotopic yields for the electromagnetic-induced fission of ^{238}U (left) and average neutron emission as a function of Z (right). The neutron emission is not measured directly but deduced from the missing mass when comparing fission fragments to the fissioning system.

two gaussian shapes, each one corresponding to a fission mode. In the super-long mode (SL), both fragments are initially strongly deformed, thus leading to large excitation energy in each fission fragment, which in turns lead to the evaporation of more neutrons. Conversely, in the standard mode I or II (S-I, S-II), one of the fragments (the heavy and the light one, respectively) has a spherical shape from the start, resulting in a limited excitation energy and a reduced neutron evaporation. The transition of shape in the In-Sn-Sb distributions corresponds to the transition between the fission modes.

The SL mode is dominant in Ag and Cd and still prevalent in In. But the S-I mode associated to the closed neutron shell N=82 becomes significant in Sn and totally dominates the fission in Sb and larger-Z nuclei. This very quick transition can be further observed on the distribution of the neutron multiplicity, presented in figure 5. The standard modes correspond to the formation of one undeformed, thus cold fragment, and therefore to a reduced value of $\bar{\nu}$. Conversely, after an SL-mode fission, the deformation energy is converted in excitation and finally in additional neutron emission.

4. Conclusions and perspectives

The first SOFIA experiment successfully delivered fission fragments yields of unprecedented quality on several fissioning systems, including very exotic ones. Detailed information about the various fission modes have also been obtained as well as the prompt neutron yields. These results will strongly constraint of models of fission and will feed the nuclear databases, allowing more refined simulations of reactor neutronics and safety studies. The next step will be the measurement of the yields in the fission of ^{236}U, the equivalent of the neutron-induced fission of ^{235}U : this experiment is scheduled for the 2nd half of 2014.

References

1. K.-H. Schmidt et al., Nucl. Phys. A 665 (2000) 221
2. M. Caamano et al., Physical Review C 88, 024605 (2013).
3. Rodríguez Sánchez, José Luis, "Simulation of the SOFIA experiment at GSI for studies of fission in inverse kinematics" (Master Thesis), 07/2012
4. H. Geissel et al., Nucl. Instr. Meth. B 70, 286-297 (1992)
5. A. Ebran et al., Nucl. Instr. Meth. A 18, 767-772 (2013)
6. E. Pellereau, PhD thesis, Université Paris-Sud

Present Status of Nuclear Data for Nuclear Nonproliferation

M. Igashira

Research Laboratory for Nuclear Reactors, Tokyo Institute of Technology,
2-12-1-N1-26 Ookayama, Meguro-ku, Tokyo 152-8550, Japan
iga@nr.titech.ac.jp

The present status of nuclear data was reviewed from the viewpoint of the research and development of non-destructive inspection methods for international controlled materials such as ^{235}U, ^{239}Pu, ^{241}Pu, D_2O, ^{6}LiD, and ^{9}Be. The nuclear characteristics of these materials were discussed, and neutron- and photon-reaction cross section data were reviewed. It was found that the accuracy of neutron data was enough for the research and development but that of photon data was not enough.

Keywords: nuclear nonproliferation; international controlled materials; non-destructive inspection; neutron cross section data; photon cross section data.

1. Introduction

Nuclear nonproliferation is one of the most important issues for the peaceful use of nuclear energy. In "treaty on the non-proliferation of nuclear weapons" (NPT), US, Russia, UK, French, and China are defined as the depositary governments. India, Pakistan, and Israel are non-member nations of NPT, and North Korea is a withdrawal nation. Namely, these nine countries are in the nuclear weapons club.

After the terrorism on September 11, 2001, we see the existence of terrorist organizations in a new light. In addition, it is believed that about 250 tactical nuclear weapons went missing due to the collapse of Soviet Union.

In these situations, the non-destructive inspection of hidden fissile materials and/or other international controlled materials (ICMs) at international ports and airports is very useful for the nuclear nonproliferation.

Neutrons and photons are thought to be promising probes for the non-destructive inspection. When a container or a piece of baggage is irradiated with neutrons, the fission reaction of the hidden fissile material will be observed by detecting prompt and/or delayed fission neutrons. In the case of photons, laser

Compton Scattering (LCS) gamma rays attract a great deal of attention. Since the LCS gamma rays are mono energetic and energy tunable, they can be applied to nuclear resonance fluorescence (NRF) for the non-destructive inspection. In addition, the LCS gamma rays can be applied to the photon induced fission of hidden fissile materials as similar as to the neutron induced fission. Moreover, the LCS gamma rays can be applied to the (γ,n) reaction of ICMs. Therefore, neutron- and photon-induced nuclear reaction cross section data are indispensable for the research and development of non-destructive inspection methods.

ICMs may be classified into two groups: the first priority group and the second priority group. The first priority group will contain fissile materials: ^{235}U, ^{239}Pu, ^{241}Pu, etc. The second priority group will contain other ICMs such as ^{6}LiD (hydrogen bomb material), D_2O (neutron moderator for ^{239}Pu production), ^{9}Be (neutron multiplier and reflector for ^{239}Pu production), etc.

In this contribution, the present status of the nuclear data for the afore-mentioned nuclides is described.

2. Nuclear Characteristics of ICMs

The neutron separation energies [1] of ^{235}U, ^{239}Pu, ^{241}Pu, ^{2}H, ^{6}Li, and ^{9}Be range from 1.67 MeV to 5.66 MeV (see Table 1), and are lower than those of ordinary structural elements such as Fe (> 7.65 MeV), Cr (> 7.94 MeV), Ni (> 7.82 MeV), W (> 6.19 MeV), and Pb (> 6.74 MeV). Therefore, the LCS gamma rays with the energy of about 6 MeV are useful for the detection of thesenuclides by (γ,n) reactions.

Table 1. Neutron separation energies and cross sections of important nuclides.

Nuclide	Neutron separation Energy (MeV)	Neutron cross section (b)
^{235}U	5.30	Fission at 0.025 eV: 585
		(n,2n) at 14 MeV: 0.548
^{239}Pu	5.65	Fission at 0.025 eV: 747
		(n,2n) at 14 MeV: 0.230
^{241}Pu	5.24	Fission at 0.025 eV: 1,012
		(n,2n) at 14 MeV: 0.603
^{2}H	2.22	Capture at 0.025 eV: 5.50×10^{-4}
		(n,2n) at 14 MeV: 0.177
^{6}Li	5.66	(n,t) reaction at 0.025 eV: 940
		(n,2n) at 14 MeV: 0.078
^{9}Be	1.67	Capture at 0.025 eV: 8.49×10^{-3}
		(n,2n) at 14 MeV: 0.484

The neutron cross sections [2] relevant to the inspection of ^{235}U, ^{239}Pu, ^{241}Pu, ^{2}H, ^{6}Li, and ^{9}Be are also shown on Table 1. The fission cross sectionsof ^{235}U, ^{239}Pu, and ^{241}Pu are largeat 0.025 eV. Therefore, thermal neutrons are useful for the inspection of these nuclides by (n,f) reactions. The capture cross sections of ^{2}H and ^{9}Be are very smallat 0.025 eV. Therefore, the neutron capture reaction is not useful for the inspection of these nuclides.In the case of the ^{6}Li(n,t)α reaction, its cross section at 0.025 eV is large, but no gamma ray is emitted from the reaction. On the other hand, since the neutron absorption by this reaction is strong, a neutron transmission or absorption method is thought to be useful for the inspection of ^{6}Li. The (n,2n) reaction cross sections at 14 MeV of ^{235}U, ^{239}Pu, ^{241}Pu, ^{2}H, ^{6}Li, and ^{9}Be range from 0.078 b to 0.603 b, i.e. those cross sections are not large. Therefore, (n,2n) reaction is thought to be not useful for the inspection of those nuclides.

3. Probes for Non-Destructive Inspection of ICMs

Concerning ^{235}U, ^{239}Pu, and ^{241}Pu in the first priority group, as mentioned above, neutrons are a promising probe for the inspection of those fissile nuclides, because their thermal neutron fission cross sections are large. The LCS gamma rays are another promising probe for those nuclides, because the LCS gamma rays with an energy of about 6 MeV are useful for the inspection of those nuclides by (γ,n) reactions. Moreover, the LCS gamma rays can be applied to the NRF and (γ,f) reactions of those nuclides. Of course, ordinary X rays are useful for the inspection of medium- and high-Z materials. Therefore, the combination of ordinary X rays, LCS gamma rays, and neutrons is suitable for the inspection of the fissile materials.

As for ^{2}H and ^{9}Be in the second priority group, the LCS gamma rays are a promising probe for the inspection by (γ,n) reactions, because their neutron separation energies are very small: 2.22 MeV for ^{2}H and 1.67 MeV for ^{9}Be. It is worthy to note that both ^{2}H and ^{9}Be have no bound states except for their ground states.

The neutron separation energy of ^{6}Li is 5.66 MeV, and is lower than those of ordinary structural elements, described above. Therefore, the LCS gamma rays are a promising probe for the inspection by (γ,n) reactions. Of course, the LCS gamma rays can be applies to NRF for the inspection of ^{6}Li. Neutrons are thought to be another promising probe for the inspection by neutron transmission and/or absorption, because the ^{6}Li(n,t)α reaction cross section at thermal energy is large: 940 b.

Consequently, in addition to ordinary X rays, the LCS gamma rays and neutrons are the promising probes for the non-destructive inspection of the ICMs.

4. Present Status of Nuclear Data relevant to Non-Destructive Inspection of ICMs

4.1. *Neutron cross section data of fissile materials*

Since fissile materials are very important for the research and development of nuclear reactors, their cross section data are contained in evaluated nuclear data libraries such as JENDL-4.0 [2] and ENDF/B-VII.1 [3]. As an example, the neutron fission cross sections of ^{235}U taken from JENDL-4.0 and ENDF/B-VII.1 are compared in Fig. 1. The agreement between the JENDL-4.0 data and ENDF/B-VII.1 data is very good except for the neutron energy region from 500 eV to 30 keV. It is worthy to note that the JENDL-4.0 treats the region above 500 eV as the unresolved resonance region.

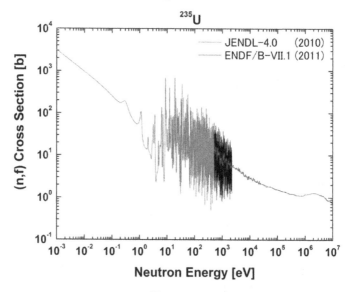

Fig. 1. Neutron fission cross sections of ^{235}U taken from the JENDL-4.0 and ENDF/B-VII.1.

The accuracy of the neutron cross section data of ^{235}U, ^{239}Pu, and ^{241}Pu is good compared to other fissile materials. As an example, the relative standard deviations of the fission cross sections of ^{239}Pu are shown in Fig. 2. Both the relative standard deviations of the JENDL-4.0 data and the ENDF/B-VII.1 data are less than 5% up to 20 MeV.

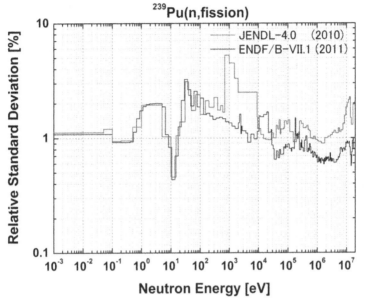

Fig. 2. Relative standard deviations for the fission cross sections of ^{239}Pu.

4.2. *Photonuclear data of ICMs*

There are photonuclear data libraries such as JENDL Photonuclear Data File 2004 [4], ENDF/B-VII Incident-Gamma Data [5], and IAEA Photonuclear Data Library [6]. However, the data contained in these libraries are mainly those above the neutron separation energy. Moreover, the accuracy of photonuclear data is not good compared to that of neutron data, because recent measurements of the photonuclear reaction cross sections are very scarce. As an example, the gamma-ray absorption cross sections of ^{2}H are shown in Fig. 3, which is taken from Ref. [6]. As shown in Fig. 3, the experimental data are very scarce below 5 MeV, and those around 20 MeV scatter around the evaluated values in the JENDL Photo Nuclear Data File 2004.

Concerning the NRF data, there is no compilation at the moment although a compilation of the NRF data was proposed [7]. Therefore, individual researchers who want to develop a non-destructive inspection method with NRF have to take information on the NRF data from individual papers. Moreover, most of NRF data are qualitative : only resonance energies and relative intensities.

232

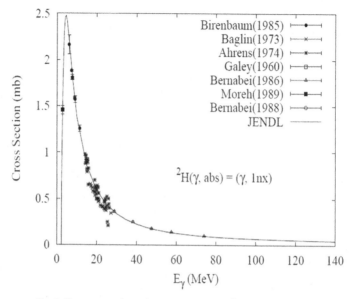

Fig. 3. Gamma-ray absorption cross sections of ^{2}H taken from ref. 6.

5. Conclusion

The present status of nuclear data was reviewed from the viewpoint of the research and development of non-destructive inspection methods for international controlled materials such as ^{235}U, ^{239}Pu, ^{241}Pu, D_2O, ^{6}LiD, and ^{9}Be. As a result, it was found that the accuracy of the neutron data was enough for the research and development but the accuracy of the photon data was not enough. In particular, it was found that there is no good compilation of the NRF data at the present moment.

Acknowledgments

The author would like to thank M. Mizumoto, K. Hara, and K. Terada for their help in this work.

References

1. R. B. Firestone *et al.*, Table of Isotopes, 8th edn.(John Wiley & Sons, 1996).
2. K. Shibata *et al.*, JENDL-4.0: A new library for nuclear science and technology, *J. Nucl. Sci. Technol.*, **48**, 1 (2011).

3. M. B. Chadwick *et al.*, ENDF/B-VII.1 Nuclear Data for Science and Technology: Cross Sections, Covariances, Fission Product Yields and Decay Data, *Nucl. Data Sheets*, **112**, 2887 (2011).

4. N. Kishida *et al.*, JENDL Photonuclear Data File, in *Proc. Int. Conf. on Nuclear Data for Science and Technology (ND2004)*, (Santa Fe, USA), Vol. **1**, p. 199 (2004).

5. M. B. Chadwick *et al.*, ENDF/B-VII.0: Next Generation Evaluated Nuclear Data Library for Nuclear Science and Technology, *Nucl. Data Sheets*, **102**, 2931 (2006).

6. IAEA-TECDOC-1178, Handbook of photonuclear data for application (2000).

7. IAEA Memo CP-D/703 (2011), IAEA Memo CP-D/721 (2013).

Development of the Experimental Photo-Nuclear Reaction Database in Hokkaido University

A. Makinaga

Faculty of Science, Hokkaido University,
Sapporo, 060-0810, Japan
makinaga@nucl.sci.hokudai.ac.jp

Nuclear databases are important tools to apply nuclear phenomena to various fields of nuclear engineering. It is now recognized that the databases must be further developed for photo-nuclear reaction data for nuclear security, safety and nonproliferation applications. Hokkaido University Nuclear Reaction Data Centre (JCPRG) has contributed to the Experimental Nuclear Reaction Data Library (EXFOR) which is developed by the International Network of Nuclear Reaction Data Centres under coordination by IAEA. We report here on the recent compilation of the nuclear data files for the photonuclear reaction.

Keywords: Nuclear data; photonuclear reaction; EXFOR.

1. Introduction

In recent years, an attractive attention is drawn to photo-nuclear reactions. Since the nuclear resonance fluorescence (NRF) below the particle threshold can acquire specific information for various kinds of nuclei, without changing the nucleus itself, there has been a great hope to improve in the nondestructive technique for nuclear security, and nuclear nonproliferation. On the other hand, in the recent nuclear physics field, NRF data can be obtained by photo-nuclear experiments using laser inverse-Compton γ-rays, or bremsstrahlung radiation with γ-ray cascade models. In order to utilize these basic data to a practical level, it is necessary to compile these experimental data in a database and construct a system that will enable users around the world to share information. The Hokkaido University Nuclear Reaction Data Centre (JCPRG)[1] has participated in developing of the nuclear reaction experiment database (EXFOR)[2], as a member of the International Network Reaction Data Centres (NRDC)[3] under the auspices of the International Atomic Energy Agency (IAEA)[4]. At present, JCPRG takes care of the experimental data for charged particle reactions and photo-nuclear reactions which are performed in Japanese experimental facilities. In this report, we introduce the international cooperation organization concerning the EXFOR maintenance and the current status of the database in Chapter 2. In Chapter 3, we introduce an overview of the EXFOR compilation procedure. In Chapter 4, the present status of NRF experimental data in the EXFOR is introduced.

2. Nuclear reaction experiment database EXFOR

The International Nuclear Reaction Data Centre consists of 14 nuclear data centres. Because of a historical reason, 4 core centres (US National Nuclear Data Center, OECD NEA Data Bank, IAEA Nuclear Data Section, Russian Nuclear Data Center) are designated as core centre. In addition, 10 specialized centres establish the collaborative relation of this NRDC activities. Table 1 shows the current NRDC member centres. The NRDC has the following tasks. (1) The construction of a bibliographic database (CINDA), (2) the construction of a nuclear reaction experiment database (EXFOR), (3) the collection of evaluated nuclear data libraries (ENDF, JENDL, etc.), (4) sharing of nuclear reaction information in the world, and (5) the construction of data file formats.

Here, we presents the EXFOR database activities. The EXFOR

Table 1. International Network of Nuclear Reaction Data Centres (NRDC) and their EXFOR compilation responsibility.

Abbreviation	Nuclear Data Centre	Responsibility
ATOMKI	ATOMKI Charged Particle Nuclear Reaction Data Centre	CPND
CAJaD	Russian Nuclear Structure and Reaction Data Centre	CPND
CDFE	Centre for Photo-nuclear Experiments Data	PhND
CJD	Russian Nuclear Data Centre	N
CNDC	China Nuclear Data Center	N,CPND
CNPD	Center of Nuclear Physics Data	CPND
JCPRG	Hokkaido University Nuclear Reaction Data Centre	CPND, PhND
NDS	IAEA Nuclear Data Section	N, CPND,PhND
NEA DB	OECD/NEA Nuclear Data Bank	N, CPND
NNDC	US National Nuclear Data Center	N, CPND, PhND
UKRNDC	Ukrainian Nuclear Data Center, Institute for Nuclear Research	N, CPND, PhND
NDPCI	Nuclear Data Physics Centre of India	N, CPND, PhND
KNDC	Korea Nuclear Data Center	N, CPND, PhND

Note: N, CPND, PhND means Neutron induced reaction, Charged particle reaction and Photo-nuclear reaction, respectively.

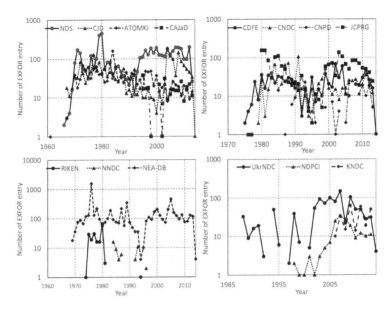

Fig. 1. The time series display for the number of EXFOR entry from each nuclear data centre. The library includes information more than 20000 experiments.

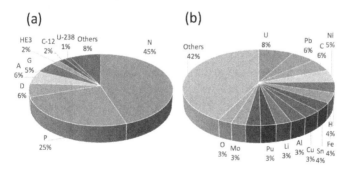

Fig. 2. The EXFOR statistics. (a) shows the percentage of incident particles in the EXFOR. (b) shows the parentage of targets in the EXFOR.

contains not only bibliographic data but also numerical data, which are obtained from authors or digitized from paper. The EXFOR has a original file unit called "entry". The present EXFOR contains more than 20000 experimental works. The number of entries recorded in the EXFOR is shown in Fig. 1 in the chronological order. In recent years, due to the active

Table 2. Main physical quantities in the EXFOR.

No.	Code	Percent	Quantity
1	CS	46.2	Cross section data
2	DAP	12.5	Partial differential data with respect to angle
3	RP	8.38	Resonance parameters
4	DA	7.95	Differential data with respect to angle
5	DAE	5.46	Differential data with respect to angle and energy
6	CSP	3.86	Partial cross section data
7	POL	3.45	Polarization data
8	FY	2.4	Fission product yields
9	RI	1.45	Resonance integrals
10	TT	1.43	Thick target yields

participation in nuclear database development activities by the Asian countries such as India, South Korea and China, it has become possible to obtain experiment information from these areas. Kazakhstan, Mongolia and other nations have recently begun participating in the EXFOR activities. Further expansion of international collaborative relationships is anticipated in future.

Figs. 2 (a) and (b) show statistics of data files stored in the EXFOR categorized by the incident particle and by the target. In contrast to neutron induced reaction data at 45% and proton induced reaction data at 25%, photo-nuclear reaction data accounts for roughly 5%. Additionally, in the classification by target, U (uranium) targets account for 8%, followed by Pb (lead) targets at 6%. With regard to the photo-nuclear reaction data, there is a additional need for intensified maintenance of older scientific publications. Table 2 shows the main physical quantities in the EXFOR. More than half of the experimental works are cross section data sets or partial differential data sets. The EXFOR also contains resonance parameters, polarization data, fission product yields and so on. At present stage, registration of electron induced reaction data is not of particularly high priority. However, this should be fully intensified in future, as similarly to NRF data. Electron induced reaction data files have also potential for adaptation to the fields of application that utilize information on nuclear properties.

3. How to compile the EXFOR

The compilation procedure of the EXFOR data file is summarized in Fig. 3. First, the NRDC network investigates bibliographic information from the published journals which include the experimental data. The main targets for registration are peer-reviewed papers and publications, although

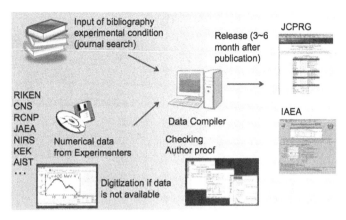

Fig. 3. EXFOR compilation procedure. Published paper is searched by IAEA and shared responsibility with the NRDC network. Basically, each nuclear data centre takes care of the compilation paper which includes the experiment performed in their own countries.

registration is also possible for conference proceedings, doctoral theses, and other items. The candidate for compilation articles are exhibited and managed via the EXFOR Compilation Control System (X4CoCoS) in IAEA. Based on the X4CoCoS, the relevant personnel at each data centre can find out which publications are assigned to their EXFOR compilation. Then, in each centre, a compilation personnel extracts the registration information, such as the experimental facility, experimental setup, experimental conditions, data reduction procedure, the way to evaluate the experimental data and experimentally obtained numerical data. Those experimental information is compiled by using editors specialized for the EXFOR compilation. The EXFOR data files prepared by an originating centre are submitted to the other centres for review, and then modified as per comments from the other center. After these procedures, EXFOR files are distributed to the nuclear data centre of each country through IAEA. Fig. 4 shows the EXFOR search system on the IAEA and JCPRG webpage.

4. Recent EXFOR compilation of the NRF data

The nuclear resonance fluorescence (NRF) is the resonant absorption of photons and de-excitation of the given excited level E_i to a level E_f in a nucleus. In case of non-overlapping resonances, cross sections for NRF can be described as $\sigma_{\gamma f}(E_i) = \sigma_\gamma(E_i)\frac{\Gamma_f}{\Gamma}$, where Γ_f is the partial width and $\Gamma = \Sigma\Gamma_f$ is the total level width. Because of the

240

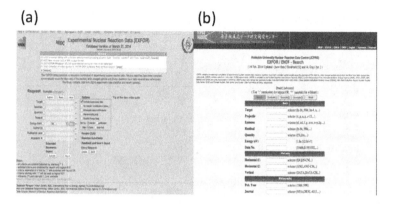

Fig. 4. (a) The EXFOR database search system in the IAEA web site. https://www-nds.iaea.org/exfor/exfor.htm (b) The EXFOR database search system in the JCPRG web site. http://www.jcprg.org/exfor/

experimental situation, NRF experiments can observe only the energy integrated cross sections $I_s = \int_0^\infty \sigma(E)dE = \frac{2J_i+1}{2J_0+1}(\frac{\pi\hbar c}{E_f})^2$. The measured intensity of γ-rays emitted to the ground state at $E_\gamma = E_i$ with an angle θ is, $I_\gamma(E_\gamma,\theta) = I_s(E_i)\Phi(E_i)\epsilon(E_\gamma)N_{at}W(\theta)\frac{\Delta\Omega}{4\pi}$, where N_{at} is number of the target nuclei per unit area, $\epsilon(E_\gamma)$ is the absolute efficiency at E_γ, $\Phi(E_i)$ is the absolute photon flux at E_i, $W(\theta)$ is the angular correlation of this transition, and $\Delta\Omega$ is the solid angle for the detector. Detailed definition of the NRF code in the EXFOR system is available in the working paper of IAEA NRDC meeting[5,6]. Table 3 shows the recent NRF experiment compiled in the EXFOR.

5. Summary

Development of the NRF database is important for the application of the nuclear engineering field. Since 2011, the NRF data is intensively compiled to the EXFOR database by NRDC. In Hokkaido University Nuclear Data Centre (JCPRG), a collaboration of EXFOR compilation for photo-nuclear data begun in 2006. JCPRG will carry out this EXFOR compilation activity continuously in future.

Acknowledgments

A. M thanks Dr. S. P. Simakov (IAEA) to provide us the information of the EXFOR compilation for the NRF data. A. M also thanks Prof. K. Kat

Table 3. Recent NRF experimental journals compiled in the EXFOR.

Entry number	Publication	Year	First author	Facility	γ-ray source	Target	E_{min} (eV)	E_{max} (eV)
G0035	J,PR/C,87,044306[7]	2013	R.Massarczyk	2GERZFK	BRST	Pt-196	4.40E+6	8.20E+06
K2408	J,PR/C,87,024301[8]	2013	T.Shizuma	2JPNAIS	LCS	Fe-56	6.93E+06	9.90E+06
L0178	J,PR/C,87,024306[9]	2013	R.Schwengner	2GERZFK,1USATNL	BRST,LCS	Kr-86	5.52E+6	1.12E+7
G0032	J,PR/C,85,014311[10]	2012	G.Schramm	2GERZFK	BRST	Se-78	4.10E+06	1.05E+07
G0034	J,PR/C,86,034313[11]	2012	N.Cooper	2GERTHD	BRST	Se-76	2.95E+06	8.71E+06
L0170	J,PR/C,85,044302[12]	2012	SL.Hammond	1USATNL	LCS	U-238	1.95E+06	4.35E+06
L0171	J,PR/C,85,044605[13]	2012	C.W.Arnold	1USATNL	LCS	Be-9	1.73E+06	5.59E+06
L0176	J,PR/C,86,034307[14]	2012	B.J.Quiter	1USAMIT	BRST	Pu-240	2.16E+06	2.74E+06
L0177	J,PR/C,86,044614[15]	2012	L.S.Myers	1USATNL	LCS	Li-6, O-16	6.00E+07	
L0182	J,PR/C,86,051302[16]	2012	C.T.Angell	1USATNL,2JPNAIS	LCS,LCS	Nd-142	4.13E+06	8.71E+06
L0183	J,PR/C,86,014319[17]	2012	R.Massarczyk	2GERZFK	BRST	Ba-136	2.98E+06	1.08E+07
G0029	J,PR/C,84,024326[18]	2011	D.Savran	2GERTHD	BRST	Xe-136	3.63E+06	8.09E+06
L0159	J,PR/C,83,034615[19]	2011	A.S.Adekola	1USATNL	LCS	Th-232	2.00E+06	4.00E+06
L0161	J,PR/C,83,041601[20]	2011	E.Kwan	1USATNL	LCS	U-235	1.66E+06	2.75E+06
L0160	J,NIM/B,269,1130[21]	2011	B.J.Quiter	1USAMIT	BRST	U-238	2.00E+06	2.47E+06
M0804	J,NIM/A,618,160[22]	2010	O.Yevetska	2GERTHD	BRST	H-1	3.25E+07	5.14E+07
L0151	J,PRL,104,072501[23]	2010	A.P.Tonchev	1USATNL	LCS	Ba-138	4.00E+06	9.10E+06
G0021	J,PR/C,82,024314[24]	2010	A.Makinaga	2GERFZK	BRST	La-139	5.50E+06	9.00E+06
G0024	J,PR/C,81,044309[25]	2010	O.Yevetska	2GERTHD	BRST	U-235	3.50E+06	4.40E+06
L0155	J,PR/C,82,054310[26]	2010	C.T.Angell	1USAMIT	BRST	Np-237	7.70E+06	1.12E+07
L0146	J,PR/C,79,037303[27]	2009	R.Schwengner	2GERFZK	BRST	MG-26	1.51E+06	1.20E+07
L0140	J,PR/C,79,014303[28]	2009	N.Benouaret	2GERFZK	BRST	Y-89		
L0147	J,PR/C,79,061302[29]	2009	G.Rusev	2GERFZK	BRST	Mo-92,94,96,98,100	4.10E+06	8.10E+06

Note: Facility Code: 2GERFZK (Helmholtz Zentrum Dresden Rossendorf in Germany), 2JPNAIS (National Institute of Advanced industrial Science and Technology in Japan), 1USATNL (Triangle Universities Nuclear Laboratory in USA) ,2GERTHD (Darmstadt University of Technology in Germany),1USAMIT (Massachusetts Institute of Technology in USA).

Source Code:BRST (Bremsstrahlung photon), LCS (Laser Inverse Compton Scattering photon)

E_{min} and E_{max} are the minimum incident energy and the maximum incident energy, respectively.

ō (Hokkaido Univ.) and Prof. M. Aikawa (Hokkaido Univ.), and N. Otuka (IAEA) to their encouragement for the EXFOR compilation.

References

1. Hokkaido University Nuclear Reaction Database Development Centre JCPRG, http://www.jcprg.org/.
2. Experimental Nuclear Reaction Data EXFOR, https://www-nds. iaea.org/exfor/exfor.htm
3. International Network of Nuclear Reaction Data Centres NRDC, https://www-nds.iaea.org/nrdc/
4. International Atomic Energy Agency, Nuclear Data Services, https: //www-nds.iaea.org/
5. S. Simakov, et al., Working paper WP2011-14, NRDC2011.
6. N. Otsuka, et al., Working paper WP2012-11, NRDC2012.
7. R. Massarczyk, et al., Phys. Rev. C **87**, 044306 (2013).
8. T. Shizuma, et al., Phys. Rev. C **87**, 024301 (2013).
9. R. Schwengner, et al., Phys. Rev. C **87**, 024306 (2013).
10. G. Schramm, et al., Phys. Rev. C **85**, 014311(2012).
11. N. Cooper, et al., Phys. Rev. C **86**, 034313 (2012).
12. SL. Hammond, et al., Phys. Rev. C **85**, 044302 (2012).
13. C. W. Arnold, et al., Phys. Rev. C **85**, 044605 (2012).
14. B. J. Quiter, et al., Phys. Rev. C **86**, 034307 (2012).
15. L. S. Myers, et al., Phys. Rev. C **086**, 044614 (2012).
16. C. T. Angell, et al., Phys. Rev. C **86**, 051302 (2012).
17. R. Massarczyk, et al., Phys. Rev. C **86**, 014319 (2012).
18. D. Savran, et al., Phys. Rev. C **84**, 024326 (2011).
19. A. S. Adekola, et al., Phys. Rev. C **83**, 034615 (2011).
20. E. Kwan, et al., Phys. Rev. C **83**, 041601 (2011).
21. B. J. Quiter, et al., Nucl. Instr. Meth. B **269**, 1130 (2011).
25. O. Yevetska, et al., Nucl. Instr. Meth. A **618**, 160 (2010).
23. A. P. Tonchev, et al., Phys. Rev. Lett. **104**, 072501 (2010).
24. A. Makinaga, et al., Phys. Rev. C **82**, 024314 (2010).
25. O. Yevetska, et al., Phys. Rev. C **81**, 044309 (2010).
26. C. T. Angell, et al., Phys. Rev. C **82**, 054310 (2010).
27. R. Schwengner, et al., Phys. Rev. C **79**, 037303 (2009).
28. N. Benouaret, et al., Phys. Rev. C **79**, 014303 (2009).
29. G. Rusev, et al., Phys. Rev. C **79**, 061302 (2009).

Laser Compton Scattering Photon Beams and Other Gamma-Ray Sources: Project for Coherent Gamma-Ray Source on Basis of Femtosecond Laser At ILC MSU *

V. G. Nedorezov

Institute for Nuclear Research RAS,
Moscow, Russia
vladimir@cpc.inr.ac.ru

A. B. Savel'ev

Lomonosov State University,
Moscow, Russia
abst@physics.msu.ru

Laser Compton back scattering photon beams and other gamma-ray sources are discussed in frame of the nuclear nonproliferation problem. New facility of ILC MSU (International Laser Center of Lomonosov Moscow State University) is described. Measured characteristics of the electron and gamma radiation in dependence on the laser parameters including the peak power, pulse duration and others are presented.

Keywords: gamma sources; laser Compton back scattering.

1. Compton back scattering

1.1. *History*

First inverse Compton scattering method was independently proposed in 1963 by Arutunyan [1] and Milburn [2], who calculated the basic characteristics of the beam, resulting in the collision of laser photons with high energy electrons. This method was experimentally confirmed in FIAN [3] and studied in detail in Frascati [4]. Widespread use of inverse Compton scattering in photonuclear experiments began in 1994 in Novosibirsk, where a series of studies of the nuclear photo-absorption and photofission was carried out [5]. Then, such installations appeared at BNL LEGS [6], Grenoble GRAAL [7], Osaka LEPS [8], Duke University Higs [9], NewSUBARU, Japan [10] and others. First femtosecond driven Compton gamma beam was created at LBNL [11].

* This work is supported by RFBR, grants 11-02-00286, 13-02-00337.

Compton beam installations that existed before 2000 are listed in Table 1, which cannot be considered to be complete. Reference herein selected first publication specified facilities, which appeared in the literature. More information on this topic can be found in [12]. Enumerate all the settings that appeared in recent years, is not possible because of their large number. The new generation Compton installations are based on the use of high-power pulsed femtosecond lasers (see, for example, [13]).

Initially, all the Compton gamma beams were obtained at the electron storage rings, because the high current of electrons (up to several hundred mA) due to their high repetition rate. High beam quality is meant not only a hard spectrum with a sufficiently high maximal energy, a high degree of polarization, but also low background.

Table 1. History of the Compton beams

First publication, year	Facility	Field of research
1964 [3]	Moscow, Lebedev PI	First experimental evidence
1976 [4]	Frascati, INFN LADON	Photonuclear reactions
1984 [5]	Novosibirsk, INP, ROKK	Photofission
1988 [6]	Brookhaven, BNL, LEGS	Meson photoproduction
1995 [7]	Grenoble, GRAAL	Heavy meson photoproduction
1998 [8]	Osaka, LEPS	Same
1997 [9]	Duke, HIgS	Photonuclear reactions
2007 [10]	SPring-8, NewSUBARU	Same
NEW HISTORY :	femtosecond laser Compton X-ray beam	
2010 [11]	LBNL, USA	Phase contrast X-ray imaging

The intensity of Compton gamma - beam is typically low (less than 10^7 photons/sec.) This limit is associated with the fact that laser photons can decrease the lifetime of the electron beam. Increasing the intensity of the Compton gamma beam is possible if you use the long-wavelength lasers. In this case, the electron energy loss are relatively small, and therefore the scattered electron can be saved on the equilibrium orbit. In such case there are no principal limitations on the intensity of the Compton beam. The first successful experimental results were obtained in this direction in Japan at several electron storage rings using long wave CO_2 lasers [10], as well at Duke University with the help of free electron laser [9].

The emergence of a new generation of gamma-ray sources based on Compton technique using femtosecond lasers was due to the possibility of their wide applications, primarily for the global issue of nuclear non-proliferation and security. The key role played the fact that new sources are very compact devices and can be installed on mobile installations.

1. 2 *Compton method*

Differential cross section of the Compton back scattering of laser photons by electrons in the laboratory frame (without polarization) according to calculations [1,2] can be represented as:

$$\frac{d\sigma}{dn} = 4\pi r_0^2 \left[\frac{K}{1+n} + \frac{1+n^2}{K} - \frac{4n^2}{(1+n^2)^2} \right] \tag{1}$$

where $K = 1 + n + \lambda$, $n = \theta \gamma$, $\gamma = E_e/m_e$, $\lambda = 2\gamma\omega/E_e$, ω – laser photon energy, θ - gamma emission angle relatively electron momentum. The gamma spectrum presented in Figure 1 is described by the equation:

$$\frac{d\sigma_0}{dE_\gamma} = \frac{\pi r_e^2}{2} \frac{m_e^2}{\omega E_2} \left[\frac{m_e^4}{(4\omega^2 - E_\gamma^2)^2} \frac{E_\gamma}{E_e - E_\gamma} - \frac{m_e^2}{\omega E_e} \frac{E\gamma}{E_e - E_\gamma} + \frac{E_e - E_\gamma}{E_\gamma} + \frac{E_e}{E_e - E_\gamma} \right] \tag{2}$$

Compton gamma energy is uniquely associated with the scattering angle:

$$E_\gamma = 4\gamma^2 \frac{\omega}{1 + n^2 + \lambda} \tag{3}$$

Figure 1 shows the calculated energy spectrum and polarization of the Compton photons. It is seen that at the maximum energy, which corresponds to a photon scattering angle of 180^0, the degree of polarization is 100%. It should be noted that until now the direct experiments to measure the polarization of gamma - quanta of medium energies have not been done yet.

The above formulas and figure 1 show that most of the intensity of the photon beam is concentrated within a small angle, which is characterized by a relativistic factor. For most existing installations electron energy of a few GeV, the characteristic angle is less than 1 mrad.

The above basic characteristics of Compton gamma beams are useful for comparison with gamma sources of new generation based on the use of pulsed femtosecond lasers.

First, it is obvious that for hard photons we need sufficiently high energy of the accelerated electrons. Thus, according to formula (3) using a laser with a wavelength of 0.5 microns and electron energy of 250 MeV, gamma beam upper limit will be about 2.6 MeV. Such energy electrons and photons respectively now reached at very high power lasers.

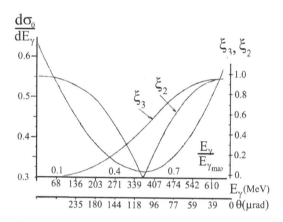

Fig. 1. Calculated spectrum $d\sigma_0/dE_\gamma$ (left scale) and polarization ξ_3 - linear ξ_2 - circular (right scale)) for inverse Compton radiation. Two abscissas show the energy (upper scale) and the angle of the emitted gamma – quanta, respectively.

2. Femtosecond laser as a gamma source

2.1. New possibilities and motivations

The development of the high power ultra-short pulse lasers within the last decade led to the extensive investigation of the laser-driven plasma as a source of high energy electrons and gamma radiation (see, for example [13]). Physics of the fast particle generation, in particular at high intensities (10^{18} W/cm^2 and above) is complicated and involves a wide variety of different processes.

The basic phenomena are still under detailed study, and one of the key plasma features that strongly impacts onto these processes is the initial electron density extent, defined mainly by the contrast of the heating pulse.

We have studied the acceleration of hot electrons with three different and typical temporal contrasts: (i) strong, above the ablation threshold, nanosecond ASE (Amplified Spontaneous Emission) accompanied by a few strong femtosecond pulses ahead of the main pulse; (ii) plasma creating femtosecond pre-pulse positioned 12 ps ahead of the main pulse, and the ASE below the ablation threshold; and (iii) the same but having additionally strong femtosecond pre-pulse 12 ns ahead of the main pulse. This enabled us to create pre-plasma with different spatial extent and density. The data were obtained with two laser pulse durations (45 and 350 fs). The motivation was to get the maximal energy of accelerated electrons and respectively gammas at relatively low initial intensity of the laser photons.

2.2. Experimental setup and results

In our study, we used radiation at a central wavelength λ of 805 nm from the 10 Hz Ti:Sapphire laser system at ILC MSU with the pulse duration $\tau_L = 45\pm5$ fs and energy after compression up to 50 mJ per pulse. The p-polarized radiation was focused by the off-axis parabola ($F/D=5$) onto the surface of a flat solid plate made of steel (mainly consisting of iron) or lead at the incidence angle of 45 degrees (Fig. 2). We can increase the pulse duration up to ~500 fs by detuning the compressor of the laser (giving the pulse with a positive or a negative chirp).

Taking into account the beam quality parameter $M^2=1.4$ and the pulse energy up to 10 mJ at the target the highest estimated vacuum peak intensity was 2×10^{18} W/cm^2 and can be lowered down to 8×10^{16} by elongation and attenuation of the laser pulse.

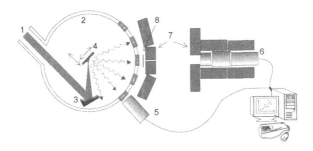

Fig. 2. The schematic view of the setup: 1 – laser radiation, 2 – vacuum chamber, 3 – off-axis parabola, 4 – target, 5&6 – scintillation detectors, 7 – lead collimators and blocks, 8 – metal foil filters.

We used two different amplification schemes in our laser system— regenerative and multipass amplifiers or two consecutive multipass amplifiers. In the first case measurements using a third order autocorrelation technique revealed that at the picoseconds time scale ASE pedestal level was ~10^{-5} relative to the peak laser intensity and a few femtosecond pre-pulses were above this level (see Fig. 3). The most intense one with amplitude of 5×10^{-3} comes ahead of the main pulse by 25 ps.

248

The second scheme (with two multipass amplifiers) provides for the much higher contrast of the laser pulse. The only one pre-pulse with amplitude of 6×10^{-7} remained (advancing the main pulse by 12 ps) at the background of the ASE pedestal having $<10^{-8}$ amplitude. It is important to note that not all the peaks seen in the 3rd correlation function are real pre-pulses indicated in the Fig. 2 by arrows and originated from the nonlinear transformation of post-pulses in the stretcher-amplifier-compressor system.

Two NaI(Tl) scintillation detectors measured the hard X-ray yield from plasma. One device (with crystal 5 mm thick) controlled the integrated yield above 5 keV in each laser shot. More details can be found elsewhere [14].

Measurements of the integrated X-ray yield from plasma in 0.1-10 MeV range showed that for the type 1 experiments it has prominent growth with pulse duration with smooth maximum for the pulse durations of the order of 300 – 500 fs, and it is almost independent on the chirp sign.

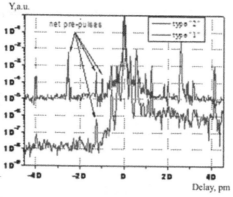

Fig.3. Third order correlation function of the "type 1" and "type 2" pulses (see text). The net pre-pulses are indicated by arrows.

Gamma spectra were measured at $\tau_L=350\pm50$ fs with the positively and the negatively chirped pulses for the steel and the lead targets. This duration corresponds to the peak intensity at the target $\sim10^{17}$ W/cm^2. The temperature grew appreciably compared to the case of the shortest pulse duration (i.e. at 10 times higher intensity): up to 255 ± 20 and 580 ± 55 keV for the steel and the lead, respectively (see Figs. 4a,b).

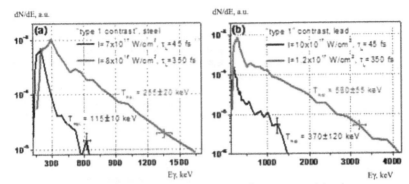

Fig.4. Hard X-ray spectra, measured for different targets, pulse contrasts and durations.

As an objective for experiments with higher (ultrarelativistic) intensity above 10^{20} W/cm^2 the proposed way of laser-plasma coupling increase may find application. Onto the modern laser systems the peak-to-pedestal ratio of 10^8–10^9 is easily accessible without complicated setup of contrast cleaning. Hence the intensity of the pedestal may be naturally obtained on the level of 10^{12} W/cm^2.

Anyway, the described in the paper results are obtained in some limited range of parameters. Further and more detailed investigation is on the way, and the research results will be published soon.

3. Other gamma ray sources

Currently known and tested different gamma sources of different energy medium types based on the different physical processes, including bremsstrahlung, positron annihilation in flight, channeling in crystals, which are described in the literature (see, for example, the monograph [15]). Recently, indications for observation of powerful electron fluxes at lightning discharges (Ground-based observations of thunderstorm-correlated fluxes of high-energy electrons, gamma rays, and neutrons), which stimulated laboratory studies high-voltage discharges. Figure 5 shows the spectrum of electrons and gamma rays from lightning discharges [16].

250

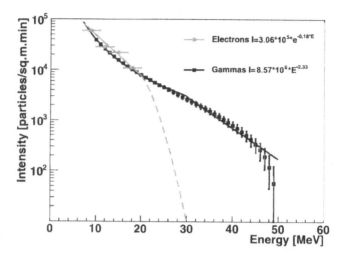

Fig. 5. Unfold electron and gamma spectra fitted by exponential and power functions [16].

The first laboratory observations of neutron bursts in discharges with crystalline deuterated targets were held recently in Ref. [17]. For the first time the emission of neutron bursts in the process of high-voltage discharge in air was observed. Experiments were carried out at average electric field strength of ~1 MV/m and discharge current of ~ 10 kA.

Two independent methods (CR-39 track detectors and plastic scintillation detectors) registered neutrons within the range from thermal energies up to energies above 10 MeV and with an average flux density of $10^6/cm^2$ per shot inside the discharge zone. Neutron generation occurs at the initial phase of the discharge and correlates with x-ray generation. The data obtained allow assuming that during the discharge fast neutrons are mainly produced.

4. Traditional methods of detection of fissile materials in containers.

The most effective and reliable method to determine fissile materials in containers today is, in our view, registration of delayed neutrons produced by the action of powerful gamma beam [18]. However, a powerful gamma beam is difficult to create in large dimensions, and secondly, it is dangerous.

Therefore we propose a two-stage procedure for scanning (see Fig. 6). First, a wide linear dimension of low-intensity beam (10^3 photons/s.pixel) irradiates the moving container as whole. Pixel size (about 1 cm) is determined by the coordinate detector, such as a scintillator viewed by a silicon photomultiplier.

By using the absorption method thus can determine the presence of heavy density substance inside the container.

In the event of such an object is held the second gamma irradiation narrow directional beam great opportunities, which is necessary to obtain a sufficiently large output delayed neutrons.

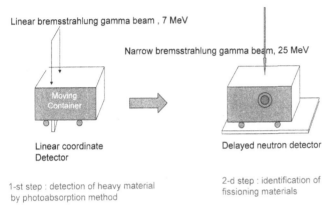

Fig. 6. *Two step method to measure the fissile material in containers.*

Numerical valuations have been done for ^{239}Pu (density $\rho=19.84$ g/cm^3), which is characterized by a minimal yield of delayed neutrons ($Y_{\gamma n} = 3.6 \cdot 10^{-3}$ per a fission) in comparison with other actinides. Object 1 cm^3 (in volume) is located in center of the container (1 m^3 in volume) with steel walls of 2 mm thickness. In order to detect 1000 delayed neutrons in the second time scale one needs an γ-ray flux of 10^{10} γ/cm^2 with the energy of 12 to 16 MeV. The appropriate intensity of bremsstrahlung beam (10^9 γ/cm$^2 \cdot$s) can be achieved using the conventional accelerators employing up to 100 μA electron current when the tungsten or tantalum target 100 micron thick is used. A detector of large size shall be used for neutron detection, most likely it will be a liquid scintillation detector used typically in the neutrino experiments. The detector sensitive area being 2 by 2 m^2 covers the solid angle about 30% of 4π. We emphasize that measurement time would in order of 10 seconds after 1 second of beam exposition. In such case about 10 background events will be detected during the measurement time.

5. Conclusion remarks

Described in this paper project for gamma source at ILC MSU is still far from its final decision. This requires both new theoretical and experimental studies.

The problem is interdisciplinary and combines specialists in nuclear physics, engineering, laser and accelerator physics. Among the priorities the choice of optimal parameters settings (design target contrast (temporary structure) beam geometry, etc.) for gamma - ray beam with a maximum energy, high angular and energy resolution can be mentioned. Preliminary measurements of the gamma spectra from metal targets on this laser showed the possibility of accelerating electrons to 5 MeV, which is not the limit yet. Nearest specific task is to create a magnetic spectrometer to measure the spectra and the outputs of the accelerated electrons under different initial conditions. Laser power will be of 10^{19} W/cm^2. Ongoing work should lead to an increase in contrast to 10^{-10} or higher with increasing energy pulse to 100 mJ. The targets of massive metal and dielectric materials, and the thin aluminum and mylar films will be explored.

Speaking about the nuclear non-proliferation and security problem, we can not exclude the traditional methods based on the use of existing accelerators like microtrons and photonuclear detection techniques. Primarily this includes registration of delayed photo neutrons from fissile material in containers. Obviously, only the combined study of different methods will be able to fulfill the task.

Acknowledgements

We thanks our colleagues Drs. G.Solodukhov and A.Turinge for useful discussions, technical editor E.Tikhonova for the manuscript preparation.

Bibliography

1. F. Arutunyan, V. Tumanyan. Compton effect on relativistic electrons and possibility of hard energy gamma beams JETF **44** 6, 2100 (1963).
2. R. H. Milburn, Electron scattering by an intense polarized photon field, Phys. Rev. Lett. **10,** 3, 75 (1963).
3. L. Ya. Kolesnikov. Quasimonohromatic and polarized high energy photon beams. *INR Int. Workshop on Electromagnetic interactions of nuclei (EMIN"75), Moscow (1975)* Publ. Nauka (1979) Moscow, 338.
4. M. P. De Pascale, G. Gindano, G. Matone e.a. The LADON Facility at Frascati, *4 course of the Int.School on Intermediate Energies., San-Miniato, (1983),* World Scientific 412 (1983).
5. A. A. Kazakov, G. Ya. Kezerashvili, L.E. Lazareva, V.G. Nedorezov, A. N. Skrinsky, G. M. Tumaikin, Yu. M. Shatunov. ROKK – back scattered Compton gamma beam at the storage ring VEPP-3, *II Int. Seminar on Spin*

Phenomena in High Energy Physics, Serpukhov (1985) 140; *IX Conf. of Charged Particles, Dubna (1985)* v.2, 268.

6. D.H. Dowell, B. Fineman, G. Giordano e.a. Laser Electron Gamma Source (LEGS) at BNL *Progress Rep. (1985) BNL 37623* , p. 29.

7. J.P.Bocquet e.a. Grenoble Accelerateur Aneeau Laser (GRAAL) at ESRF. *XIII Particle an Nuclear Int. Conf., (PANIC)* Perudja, (1993).

8. M. Fujiwara, T. Hotta, T. Kinashi e.a. Laser Electron Photon Source (LEPS) at SP-ring 8. Acta Physica Polonica B **29** (1998) 141-156.

9. V.Litvinenko e.a. Gamma-Ray Production in a Storage Ring Free-Electron Laser, Phys. Rev. Lett. **78**, 4569-4572 (1997).

10. S. Miyamoto e.a. Laser Compton Back-scattering Gamma-ray Beam-line on New SUBARU, Radiation measurements, **41**, S179-S185 (2007).

11. S. Kneip, C. McGuffey, F. Dollar, M. S. Bloom, V. Chvykov et al. X-ray phase contrast imaging of biological specimens with femtosecond pulses of betatron radiation from a compact laser plasma wakefield. Accelerator, Appl. Phys. Lett. **99**, 093701 (2011).

12. V. G. Nedorezov, A. A. Turinge, Yu. M. Shatunov. Photonuclear experiments using back scattering photon beams. Uspekhi Fiz. Nauk, **174**, 4 354 – 370 (2004).

13. A. V. Andreev, V. M. Gordienko, A. B. Savel'ev. Nuclear processes in the high temperature induced by extra short laser pulse. Quantum electronics, **31**, 8, 941-956 (2001).

14. K. a. Ivanov, S. a. Shulyapov, a. a. Turinge e.a. X-Ray Diagnostics of Ultrashort Laser-Driven Plasma: Experiment and Simulations, *Contributions to Plasma Physics* **53**(2), 116–121 (2013).

15. V. G. Nedorezov, Yu. N. Ranyuk. Photofission above the giant resonance, *Naukova Dumka, Kiev, ISBN 5-12-000869-0* (1989).

16. A. Chilingarian e.a. Ground-based observations of thunderstorm-correlated fluxes of high-energy electrons, gamma rays, and neutrons. Phys. Rev. **D 82**, 043009 (2010).

17. V. G. Nedorezov e.a. Identification of fissile materials in the railway and sea containers. *INR PNLAB- INR rep.* , p/41-46, (2006).

18. Л. V. Agafonov, A. V. Bagulya, O. D. Dalkarov e.a. Observation of Neutron Bursts Produced by Laboratory High-Voltage Atmospheric Discharge, Phys. Rev. Lett. **111**, 115003(2013).

Laser Driven Ion Acceleration Study in JAEA

K. Kondo

Quantum Beam Science Center, Japan Atomic Energy Agency,
8-1-7 Umemidai, Kizugawa, Kyoto, 617-0215, Japan
kondo.kiminori@jaea.go.jp
wwwapr.kansai.jaea.go.jp/aprc/

Using a high-contrast (10^{10}:1) and high-intensity (10^{21} W/cm^2) laser pulse with a duration of 40 fs from an optical parametric chirped-pulse amplification/Ti:sapphire hybrid laser system in JAEA, a 40 MeV proton bunch is obtained, which is a record for laser pulse with energy less than 10 J. The efficiency for generation of protons with kinetic energy above 15 MeV is 0.1%.

Keywords: Laser driven ion acceleration; chirped-pulse amplification; target normal sheath acceleration.

We have demonstrated the 40MeV proton acceleration from micrometer-scale thick metal foils irradiated by 7.5 J energy 40 fs duration 800 nm wavelength laser pulses with a temporal contrast of 10^{10}:1 focused to an intensity of 1×10^{21} W/cm^2. This is the highest proton energy for laser pulses with energy of <10 J[1]. Moreover, we have measured the generation of almost fully stripped aluminum ions of energy of over 10 MeV/u with specialized track detectors for catching high Z and middle Z ions.

The J-KAREN 200 TW optical parametric chirped-pulse amplification (OPCPA)/Ti:sap-phire hybrid laser system at the Kansai Photon Science Institute of the Japan Atomic Energy Agency (JAEA) was used. We focused the p-polarized laser pulse to the intensity of up to 1×10^{21} W/cm^2 at the incidence angle of 45°. No plasma mirror [2,3] was used to obtain the 10^{10} contrast achieved with a saturable absorber inserted between a high-energy CPA oscillator and a stretcher [4]

With an f/2 off-axis parabolic mirror, the laser pulses are focused in a spot that is 3 μm (FWHM) along the laser polarization and 2 μm in the perpendicular plane (measured in vacuum). 23% of the fulllaser energy was contained in the 3 μm× 2 μm spot.

With a cross-correlation measurement of temporal contrast at the 50 TW operation mode, a good temporal contrast of 3×10^{10}:1 has been checked to be maintained down to about 20 ps before the main pulse. At the 200 TW peak power level, we estimate the contrast to be 1×10^{10}:1, referring to the result in ref. [4]. The temporal profile of the main pulse was also optimized with an acousto-optic programmable dispersive filter (AOPDF; by FastLite) using feedback from a transient grating frequency-resolved optical gating (TG-FROG) apparatus. In this way, we have achieved the main pulse FWHM duration of 36 fs and the effective duration of 40 fs. After the optimization, the sub-ps pedestal is sup-pressed to one-hundredth before the optimization.

In addition to the proton spectral information we have also used a magnetic electron spectrometer (ESM), which was placed along the laser propagation axis, to obtain the correlated electron spectrum and slope temperature. The typical electron slope temperature (at higher energy) was 10 MeV, which corresponds to a normalized vector potential $a_0 = 0.85 \times 10^{-9}(\mathrm{I[W/cm^2]})^{1/2}\lambda[\mu m] \sim 20$ according to the ponderomotive electron temperature scaling [5]. This is consistent with the estimated interaction intensity of 1×10^{21} W/cm^2.

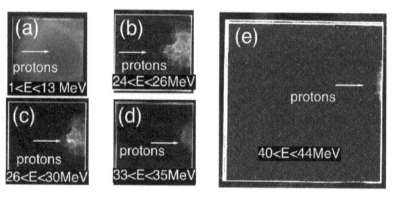

Fig. 1. Proton beam footprints on the CR39 plates for an 800 nm thick Al target irradiated at the intensity of 1×10^{21} W/cm^2 for the indicated proton energy ranges.

Energy-resolved spatial distributions of proton fluence along the target normal were determined using nuclear track detection with CR-39 films, which are sensitive to ions but not x-rays and electrons. A stack of many CR-39 plates, each of thickness 0.83 mm (by Nagase Landauer Ltd.), shielded by a 13 µm Al

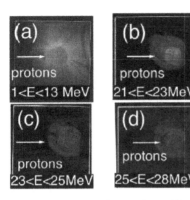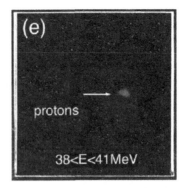

Fig. 2. Proton beam footprints on the CR39 plates for a 2 μm thick SUS target irradiated at the intensity of 1×10^{21} W/cm^2 for the indicated proton energy ranges.

filter, was placed 55 mm downstream from the foil target. Tracks formed by impinging protons are revealed by etching each plate with a 6.0 N KOH solution at 70 °C. The tracks give the ion energy and angular distribution since the ion energy determines the penetration range. The footprints consisting of etched pits generated by protons corresponding to different energy intervals with the maximum energy of about 40 MeV are shown in Figs. 1 and 2 for the 0.8 μm aluminum (Al) foil and 2 μm steel use stainless (SUS) foil, respectively. Protons originate from the water and hydrocarbon contaminants that are always present on target surfaces. We note that SUS foils arerobust and can be supplied using a tape reel for a repetitive operation. As seen in Figs. 1 and 2, the proton propagation direction is close to the target normal for relatively low proton energies and shifts toward the laser propagation direction for high energies. In the case of the Al foils, protons with energies >40 MeV deflect from the target normal by more than 20°, as is seen in Fig. 1. A thick SUS target produces a relatively small spot of protons with energies of about 38 MeV shifted from the target normal by 10°, as is seen in Fig. 2. The observed deflection of protons can be explained by bending of the target due to ablation caused by a pedestal of the laser pulse with the duration of 500 ps and intensity of 1×10^{11} W/cm^2. The resulting fluence of 50 J/cm^{2}can be large enough to displace the 0.8 μm Al foil by ~1 μm. Nearly absent (a few percent) specular reflection at 10^{20} W/cm^2 indicates a substantial absorption of the main pulse in few-micrometer plasma created by the pedestal at the target surface. For less dense targets, a proton beam deflection is also caused by a quasi-static magnetic field created by the main laser pulse, whereas for denser and thinner targets it is due to relativistic effects at oblique incidence. The proton energy spectrum is exemplified in Fig. 3

for protons emitted within unit solid angle of 1 msr from the Al foil. It is obtained by counting the total number of protons (etched pits) at each CR-39 plate. From this we can estimate a single shot conversion efficiency of the laser pulse energy to proton kinetic energy. For the Al and SUS foils, the conversion efficiency into protons with energies above 15 MeV is 0.1%, corresponding to the proton bunch energy of 7.5 mJ. The observed energy spectrum is in agreement with known scalings [6].

To estimate the generation of high Z and middle Z energetic ions in this interaction, we have inserted PET films and Kapton films in the stack for particle detection at the other experiment mentioned above. PET is sensitive for atoms heavier than helium, and Kapton is sensitive for atoms heavier than sodium. From the etch pits in this stack, the generation of aluminum, oxygen and carbon ions that were almost fully stripped have been possibly accelerated up to 12 MeV/u, 18MeV/u, and 16 MeV/u, respectively.

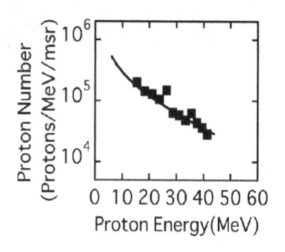

Fig. 3 Proton energy spectrum for 0.8 μm thick Al target.

In conclusion, the maximum proton energy of 40 MeV was obtained using only 7 J pulses from a Ti:sapphire laser system focused to very high intensity and with high contrast obtained without plasma mirrors. This proton energy enhancement and efficiency indicate the potential for simple thin metallic foils as sources for future integrated laser-driven ion accelerator systems. Heavy ions were also measured to be accelerated up to the almost half energy of protons. This is consistent with the mass difference between heavy ions and proton with in the detection accuracy.

References

1. K. Ogura, et al., *Opt. Lett.* **37**, 2868 (2012).
2. B. Dromey, et al., *Rev. Sci. Instrum.* **75**, 645 (2004).

3. G. Doumy, et al., *Phys. Rev. E* **69**, 026402 (2004).
4. H. Kiriyama, et al., *Opt. Lett.* **35**, 1497 (2010).
5. S. C. Wilks, et al., *Phys. Rev. Lett.* **69**, 1383 (1992).
6. M. Passoni and M. Lontano, *Phys. Rev. Lett.* **101**, 115001 (2008).

Status of New JENDL Photonuclear Data File

Kazuaki Kosako

Institute of Technology, Shimizu Corporation
4-17, Etchujima 3-Chome, Koto-ku, Tokyo 135-8530, Japan

Toru Murata

JENDL Committee, Japan Atomic Energy Agency
Tokai-mura, Naka-gun, Ibaraki-ken 319-1195, Japan

Nobuyuki Iwamoto

Nuclear Data Center, Japan Atomic Energy Agency
Tokai-mura, Naka-gun, Ibaraki-ken 319-1195, Japan

We have developed a new photonuclear data file, JENDL/PD-2014, in order to further increase the accuracy of cross sections and nuclide yields. It was evaluated by using the photoabsorption cross sections based on giant dipole resonance and the results calculated by nuclear reaction models. The number of nuclides is increased to 181 in order to take into account the use in many application fields. The photoneutron yields from some of shielding materials were examined in comparison with a experiment.

1. Introduction

Previous version of JENDL photonuclear data file, JENDL/PD-2004 [1], was released in 2004. This file contains photon-induced reaction cross section data of 68 nuclides from ^{2}H to ^{237}Np, in which ^{235}U, ^{238}U and ^{237}Np were included as fissile nuclides. LA150 [2] and KAERI [3] have also been known as photonuclear data files; the former was released with 12 nuclides in 1999, and the latter was done with 143 nuclides in 2000. However, LA150 and KAERI don't take into account the fissile nuclides.

The recent three dimensional Monte Carlo transport codes such as MCNP5 [4] and PHITS2 [5] have a function which can treat photon-induced reactions from the photonuclear data file. Owing to these Monte Carlo codes, it is expected that the use of photonuclear data spreads to various areas. However, the verification of photonuclear data is hardly considered because very few experimental data are available.

At present, photonuclear data file are used in area of the shielding of medical electron linac, the detection of the fissile materials for nuclear security and nonproliferation, and so on. The reaction of photonuclear production occurs at photon energy greater than about 5 to 10 MeV, depending on nuclides. Therefore, in the medical linac with high energy electrons from 15 to 25 MeV, there are two problems which stem from the secondary gamma-rays generated by the photoneutrons produced in the shielding wall and the radioactive waste generated by the photoneutrons produced at the linac target. The detection of fissile materials makes use of the characteristics with different threshold energies of photoneutron production and photofission reactions for every nuclide. As a result, the accuracy of photonuclear data becomes very important for the shielding design and nuclear security.

Thus, we have newly evaluated a photonuclear data file. The purpose of the present evaluation is (1) to improve cross sections from JENDL/PD-2004, some of whose data are inconsistent with measured data, (2) to systematically evaluate nuclear data on mainly light nuclides, (3) to validate nuclear data based on an experiment, and (4) to increase the number of nuclides.

2. Evaluation of Photonuclear Data

The upper limit of incident photon energy in the new JENDL photonuclear data file was fixed at 140 MeV to avoid the influence of muon production. This is the same as JENDL/PD-2004. The lowestenergy is different every nuclide and is set to the lowest particle emission threshold energy by photonuclear reaction. In this evaluation (γ,γ) and (γ,γ') reactions were not taken into account even below the particle emission threshold energy. This is because the gamma-ray emission probability is known to be quite small, compared with the particle emission ones.

For light stable nuclides of He to Ca, if the experimental cross sections of the (γ,xn) reactions compiled by Dietrich and Berman [6] and the photoabsorption reactions [7-11] were available, they were analyzed with the sum of the resonance and quasi-deuteron models [12-14]. If there were no experimental data, the resonance absorption cross sections were calculated using the giant dipole resonance (GDR) parameters given in the RIPL-3 [15]. In addition, if the GDR parameters were not provided even in the RIPL-3, the resonance structures were estimated using the level scheme given in the ENSDF [16] and the neutron resonance parameters [17].

For heavier stable and unstable nuclides up to Lr, the resonance absorption cross sections were calculated using the GDR parameters given in the RIPL-3. The GDR parameters for dominant stable nuclides were modified from the comparisons with experimental data and were applied to unstable nuclide as well as other stable ones.

The resonance formula with the GDR parameters to calculate the relative resonance absorption cross sections is

$$\sigma_{abs}(E) = \sigma_0 \sum_{i=1}^{2} \frac{a_i E^2 \Gamma_i^2}{(E^2 - E_i^2)^2 + E^2 \Gamma_i^2} , \tag{1}$$

where σ_0 is the normalization constant to satisfy the GDR sum rule [18], E is the incident photon energy, E_i and Γ_i are the energy and total width of i-th resonance, and a_i denotes the ratio of the long axis to the short axis of nuclide deformation (a_2=1).

In the quasi-deuteron model, the photo absorption cross section is given by

$$\sigma_{qd}(E) = R_{qd}(Z) \frac{NZ}{A} \left(\frac{E_b(E - E_b)}{E^2} \right)^{3/2} P(E) , \tag{2}$$

where N, Z and A are the neutron, proton and mass numbers of target nuclide, $R_{qd}(Z)$ is the coefficient to reproduce experimental data in the energy region of quasi-deuteron model (R_{qd}=159.2 for the light nuclides, R_{qd}=120 for the nuclides with $Z \geq 83$, $R_{qd}(Z) = 145 - 0.362 \times Z$ for the other nuclides), E_b is the neutron-proton binding energy of deuteron (2.223 MeV), and $P(E)$ is the Pauli blocking factor [19].

In the GDR sum rule [18], the photo absorption cross sections approximately satisfy the following sum rule relationship

$$\int_0^{140} \left(\sigma_{abs}(E) + \sigma_{qd}(E) \right) dE = C_s(Z) \frac{NZ}{A} , \tag{3}$$

where C_s is the normalization variable ($C_s = 110$ mb $\cdot$ MeV for the light nuclides, $C_s = 115$ for nuclides with $20 < Z < 84$, $C_s(Z) = 115 - 0.6666(Z - 83)$ for nuclides with $Z \geq 84$).

The photon-induced fission cross sections were obtained from the result calculated by the FISCAL2003 code [20]. The photon-induced fission was considered for nuclides heavier than Tb ($Z = 65$).

The reaction cross sections were obtained using the branching ratios calculated by the ALICE-F code [21], except for the (γ,x1n) and (γ,x2n) reactions of the light nuclides for which cross sections were measured. The branching ratios to the total reaction were calculated for the production of particles and a sufficiently wide (Z,A) range of residual nuclides. The energy distributions of outgoing particles by photonuclear reactions were also calculated by ALICE-F.

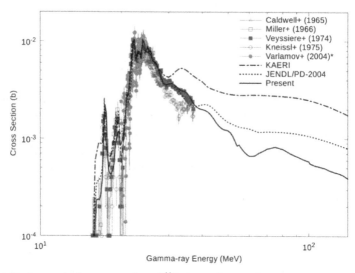

Figure 1. Neutron production cross sections of ^{16}O photonuclear reaction.

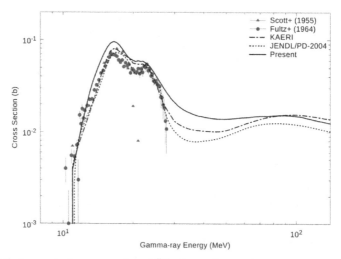

Figure 2. Neutron production cross sections of ^{63}Cu photonuclear reaction.

Figure 1 shows the evaluated neutron production cross sections and the experimental data of ^{16}O. The present evaluation is in good agreement with the experimental data. In contrast, KAERI has larger cross sections above the photon energy of 30 MeV. Figure 2 represents the evaluated neutron production cross sections and the experimental data of [63]Cu. The present data were evaluated with cross sections higher than the experimental data in the energy range of 12 to 23 MeV, based on theexperiments described in Chapter 3.

3. Comparison with Experiment

The experiment of photoneutron angular distributions was performed using the 45 MeV electron linear accelerator of Hokkaido University [22]. In this experiment, the central energies of incident electrons were 18, 28, and 38 MeV and they had the full widths at half maximum of 4, 4, and 6 MeV, respectively. The irradiated targets were aluminum, iron, copper, tungsten, and lead with 4 cm × 4 cm area. The incident electrons produced the bremsstrahlung photons in the target, and subsequently photoneutron reactions occurred in the target. The photoneutrons were measured as the neutron dose rates by two kinds of neutron dosimeters which were fabricated by Chiyoda Technol. and Nagase-Landaure and were set at angles of 30, 45, 60, 90, and 120 degrees. The measured neutron ambient dose equivalent rates (called "dose rates" for simplicity) were normalized to dose rate per unit electron beam current.

Angular distributions of photoneutrons were calculated with the MCNP5 code including the electron-photon cascade function to produce the bremsstrahlung photons. The cross section data libraries used in MCNP5 were EL3 [23] for electron-photon cascade reactions, MCPLIB04 [23] for photon interactions, and LA150 and presently evaluated data for photonuclear reactions. The dose rates were obtained from the calculated neutron flux and the flux-to-dose-rate conversion coefficients given by ICRP-74 [24].

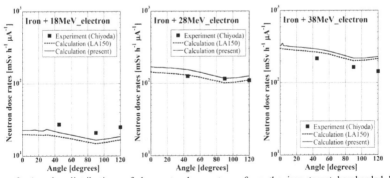

Figure 3. Angular distributions of dose rates by neutrons from the iron target bombarded by electrons of 18, 28, and 38 MeV.

266

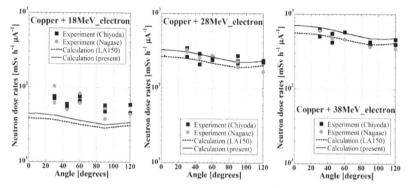

Figure 4. Angular distributions of dose rates by neutrons from the copper target bombarded by electrons of 18, 28, and 38 MeV.

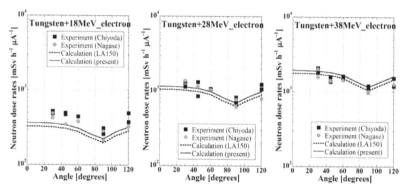

Figure 5. Angular distributions of dose rates by neutrons from the tungsten target bombarded by electrons of 18, 28, and 38 MeV.

Figures 3, 4 and 5 show the angular distributions of dose rates of neutrons emitted from the iron, copper, and tungsten targets, respectively, bombarded by 18, 28, and 38 MeV electrons. The experimental data are shown by symbols. The solid and dotted lines indicate the calculated data using the LA150 and present libraries. The two same symbols in one panel indicate the results of the repeated experiments under the same conditions. The shapes of the experimental angular distributions are almost flat and are in good agreement with the calculated ones. The minimum values are found at 90 degrees because the path of photoneutrons has the maximum length in the target. From the results of Figures 3-5, the photoneutron yields by the present data are uniformly 10-15% larger than those by LA150 at all angles. For 18 MeV electrons, the calculated dose rates using the present data give a 10-50% underestimation relative to the experimental data. For 28 and 38 MeV electrons, the calculated results are in good agreement with the experimental data. It is considered that the disagreement found in the 18 MeV case needs both reviews of the evaluated values of the photoneutron yields around 18 MeV and the experimental data.

4. Status of New JENDL Photonuclear Data File

The new JENDL photonuclear data file wasverified by comparing the evaluated data with the experimental photonuclear reaction data. The total neutron yields for aluminum, iron, copper, tungsten, and lead were also examined by the calculations.

In this new evaluation all nuclides in JENDL/PD-2004, except for ^{2}H, were revised by the methods mentioned above. The new photonuclear data file is planned to be released as JENDL/PD-2014 in this spring. The standard version stores 181 nuclides of the elements important in the field of atomic energy. The expanded version stores about 2500 nuclides including unstable nuclides up to Lr. The fissile materials are limited to three important nuclides ^{235}U, ^{238}U, and ^{237}Np in the standard version, but their considered range is extended from Tb to Lr in the expanded version. A photonuclear reaction library for MCNP5 is also prepared and is going to be released.

5. Conclusion

We have developed the new photonuclear data file in the photon energy range from particle emission threshold (typically 5-10 MeV) to 140 MeV. The systematic evaluations for light nuclides were done with the resonance and quasi-deuteron models. The evaluated data showed good agreement with the experimental data. The validation of nuclear data on structural materials was confirmed by the analysis of the experimental data. The calculated dose rates for 28 and 38 MeV electrons are in good agreement with the experimental data, while those for 18 MeV electrons give a 10-50% underestimation to the experimental data.

The release of new JENDL photonuclear data file, JENDL/PD-2014, is scheduled in this spring. The expanded version as well as the standard one of JENDL/PD-2014 will broaden the field of application.

References

1. N. Kishida, T. Murata, T. Asami, K. Kosako, K. Maki, H. Harada, Y.-O. Lee, J. Cheng, S. Chiba and T. Fukahori, *Proc. Int. Conf. on Nucl. Data for Sci. Technol.,* Santa Fe, New Mexico, USA, Sep. 26-Oct. 1, 2004, Vol. 1, 199 (2004).

2. M. B. Chadwick, LA-UR-98-4139 (1998).
3. Y. O. Lee and Y. Han, KAERI/TR-1512/2000 (2000).
4. X-5 Monte Carlo Team, LA-UR-03-1987 (2003).
5. T. Sato, K. Niita, N. Matsuda, S. Hashimoto, Y. Iwamoto, S. Noda, T. Ogawa, H. Iwase, H. Nakashima, T. Fukahori, K. Okumura, T. Kai, S. Chiba, T. Furuta and L. Sihver, *J. Nucl. Sci. Technol.* 50, 913 (2013).
6. S. S. Dietrich and B. L. Berman, *Atomic Data and Nuclear Data Tables* 38 199 (1988).
7. J. Ahrens, H. Borchert, K. H. Czock, H. B. Eppler, H. Gimm, H. Gundrum, M. Kroning, P. Riehn, G. Sita Ram, A. Zieger and B. Ziegler, *Nucl. Phys.* A251, 479 (1975).
8. J. Ahrens, *Nucl. Phys.* A 446, 229 (1985).
9. H. Harada, Y. Shigetome, H. Ohgaki, T. Noguchi and T. Yamazaki, *Phys. Rev. Lett.* 80, 33 (1998).
10. H. Harada, K. Furutaka, Y. Shigetome, H. Ohgaki and H. Toyokawa, *J. Nucl. Sci. Technol.* 35, 733 (1998).
11. H. Harada, K. Furutaka, H. Ohgaki and H. Toyokawa, *J. Nucl. Sci. Technol.* 38, 465 (2001).
12. P. M. Endt and P. B. Smith (Eds.), "Nuclear Reactions II", Chap. 3 "The Giant Resonance of the Nuclear Photoeffect", North-Holland (1962).
13. J. S. Levinger, *Phys. Rev.* 84, 43 (1951).
14. T. Murata, K. Kosako and T. Fukahori, JAERI-Conf 2011-002, 235 (2011).
15. R. Capote, M. Herman, P. Oblozinsky, P. G. Young, S. Goriely, T. Belgya, A. V. Ignatyuk, A. J. Koning, S. Hilaire, V. A. Plujko, M. Avrigeanu, O. Bersillon, M. B. Chadwick, T. Fukahori, Zhigang Ge, Yinlu Han, S. Kailas, J. Kopecky, V. M. Maslov, G. Reffo, M. Sin, E. Sh. Soukhovitskii and P Talou, *Nucl. Data Sheets* 110, 3107 (2009).
16. J. K. Tuli, BNL-NCS-51655-01/02-Rev (2001).
17. S. A. Mughabghab, "Atlas of Neutron Resonances, Fifth edition," Elsevier Science (2006).
18. A. deShalit, H. Feshbach and E. Sheldon (Rev.), "Theoretical Nuclear Physics Vol. 1: Nuclear Structure", John Wiley (1974).
19. M. B. Chadwick, P. Oblozinsky, P. E. Hodgson and G. Reffo, *Phys. Rev.* C44, 814 (1991).
20. T. Fukahori, *Proc. Int. Conf. on Nucl. Data for Sci. Technol.,* Santa Fe, New Mexico, USA, Sep. 26-Oct. 1, 2004, Vol. 1, 47 (2004).
21. T. Fukahori, JAERI-M 92-039, 114 (1992).
22. K. Kosako, K. Oishi, T. Nakamura, M. Takada, K. Sato, T. Kamiyama and Y. Kiyanagi, *J. Nucl. Sci. Technol.* 48, 227 (2011).
23. RSICC, CCC-710/MCNP (2003).
24. ICRP, ICRP Publication 74, *Ann. ICRP*, 26, Pergamon Press (1995).

Nuclear Research With E$_\gamma$≤15 MeV Photons

C. Rangacharyulu

Department of Physics and Engineering Physics,
University of Saskatchewan
Saskatoon, SK, S7N 5E2, Canada
chary.r@usask.ca

The advent of MeV photon beam facilities from the Compton scattering of laser beams off ultra-relativistic electrons at modern synchrotron light sources offers a unique tool for nuclear science and technology research. The distinct spectrum with high intensities at the endpoint energies suggests several interesting applications in addition to photo nuclear physics studies. We present some examples for photons up to 15 MeV energies.

Keywords: compton scatter; laser – electron beams; MeV photons; nuclear science and technologies.

1. Introduction

Studies of photon induced nuclear reactions have a long history, dating back to 1930s. In the past, Bremsstrahlung beams from electron accelerators along with emanations from radioactive sources were the principal sources for nuclear and particle physics research and applications. The Compton scattering of photons off a stationary electron is a well known phenomenon, taught in undergraduate courses as an example of particle characteristics of electromagnetic waves. In 1948, Feenberg and Primakoff [1], in their studies of energy exchange mechanisms of ultra relativistic electrons with photons in cosmic ray environments, showed that a) the final state photons in Compton scattering are of energies higher than the ones in initial state and b) in head-on collisions of electrons and photons, photons in the final state are mostly along the direction of electrons. Quantitatively, we can write,

$$E^{\max}_{\gamma,out} \approx 4E_{\gamma,in} \times \gamma^2_e \quad \text{and} \quad \left\langle \theta^2 \right\rangle_{\gamma,out} \approx 1/\gamma^2_e \qquad (1)$$

where γ_e is the Lorentz factor of electrons ($\gamma_e = E_e/m_e$), $E_{\gamma, in}$ and $E^{max}_{\gamma, out}$ are the energies of photons before scattering and maximum energy of outgoing photons, respectively and θ is the outgoing photon direction with respect to the flight path of electrons. The advent of high energy electron accelerators and laser beams led to the suggestion that this phenomenon can be observed in the laboratory [2,3]. The directionality of outgoing photons suggest that they can serve as secondary beams just as Bremsstrahlung, with one important useful difference. The photons due to Laser Compton scatter (LCS) are of maximum intensity at the endpoint energy $E^{max}_{\gamma, out}$, unlike Bremsstrahlung photons which are of highest intensities at lowest energies. Also, a salient feature of the LCS beams are the high degree of polarization which is not so easily achieved in Bremsstrahlung facilities.

In recent decades, the laser back scatter photon beam facilities at electron synchrotrons have become attractive alternates to conventional Bremsstrahlung beams. A good summary of the beam physics and facilities till 2000 was given in ref. 4. The 3rd generation synchrotrons with circulating electron beam currents of a few hundred milli Amperes are becoming common. Also, high power lasers of infrared and far-infrared wavelengths are easily procured at affordable prices. It is fortuitous that these laser frequency- electron energy combinations yield photons of a few MeV energies, just the right range for nuclear science and technology uses. Below, we offer a glimpse of science and technology projects that one may pursue at these facilities.

2. Photon Interactions with Matter

A look at the energy dependence of photon interactions with matter is quite informative of the suitability of a few MeV photons for nuclear research.

Figure 1 shows the experimental data of total interaction cross sections along with the theoretical cross sections of various physical processes (photo-electric, Compton, and pair production) in lead material. It is note worthy that at very low energies (a few hundred keV or less), atomic and molecular properties of matter are the sole contributors to the photon interactions. While nuclear processes become important at higher energies, there is only a narrow window of about a few tens of MeV (~20-30 MeV) where photon interactions are sensitive to nuclear structures and reaction mechanisms. Also, for energies below particle emission thresholds (8 MeV or less), properties of discrete energy levels of isotopes play important roles in photo nuclear scattering phenomena. This feature has been heavily exploited for elucidating basic nuclear physics.

Above the particle threshold, an ubiquitous feature of all nuclei is the giant dipole resonance, labeled as g.d.r., in Fig. 1. This resonance plays a very important role in the particle emissions and thus background radiations in nuclear interactions. The very significant feature of this resonance is that its location and width vary monotonously with mass number. Quantitatively, the resonance is centered at E_x (MeV) $\sim 77 \times A^{-1/3}$ and the width of resonance is Γ(MeV)= $h/\tau \sim 22xA^{-1/3}$. For masses A=100-200, the resonance lies around 14-16 MeV with widths of about 4 MeV.

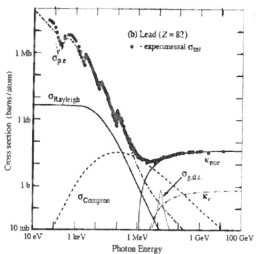

Fig. 1. Photon interaction cross sections in lead material for 10eV < Eγ < 100 GeV. Experimental data of total cross sections and model cross sections for various physical processes are shown. Figure adopted from Ref. [5]

It is quite heartening that the photons of MeV energies are within reach of the 1-3 GeV electron synchrotrons in conjunction with the industry standard lasers such as Nd:YAG and CO_2 systems of emission wavelengths about 1000 nm and 10,000 nm respectively. Below, we discuss some possible basic nuclear physics, nuclear astrophysics and technological application research at these facilities.

3. Nuclear Physics and Nuclear Astrophysics with MeV Photons

Nuclear physics with electromagnetic probe has been a very favorite topic of study for several decades. The main reason is that the interaction mechanism at the level of one photon exchange is least perturbing and is described by the well established quantum electrodynamics. While nuclear physics studies focus on structures and symmetries spanning wide range of energies, nuclear astrophysics is concerned with nucleosynthesis, occurring above the particle threshold and is probed by photons of several MeV. Groups at Osaka [6,7] have long been engaged in these studies and they continue to make extensive contributions in nuclear astrophysics. An ingenuous development is the recognition of the correspondence between the neutrino-nucleus interaction and the corresponding photon–nucleus interactions, which enables one to deduce effects of the neutral current weak interactions from the electromagnetic probe results.

3.1. *Symmetry Tests in Nuclear Physics*

Over the last few decades, nuclear physics served as a fertile laboratory for tests of fundamental symmetries of interest to both nuclear and particle physics. The physics principle of testing these symmetries is simple. It is to identify an observable which has its corresponding mathematical operation as odd (change of sign) under the symmetry operation. If the symmetry is preserved, the expectation value of the observable is zero. A non-zero expectation value signals breakdown of the symmetry. The best celebrated example is the asymmetry of beta particle emission in the decay of polarized ^{60}Co nuclei as a proof of parity violation.

3.1.1. *Test of time reversal symmetry in gamma decays*

A few experiments were motivated to examine the validity of time reversal symmetry, which is of fundamental interest for particle physics and detailed balance of nuclear reactions. In photon processes, the observables nuclear spin (**j**), photon momentum (**k**), photon spin (**s**), and its polarization (**e**) are odd under time reversal. A test of the symmetry involves design of an experiment with resultant scalar product of (2n+1) for n=1,2,... observables. Also, odd- T phase factors in multipole mixing ratios of photon cascades are of interest. It is to be stressed that the laser scatter photons at the light sources can open up a new area of study due to the high degree of polarization of the incident photon beams. Table 1 lists a few promisingcases for these studies.

Table 1. A few gamma cascades of possible interest for T-reversal symmetry tests. The data are compiled from Nuclear data tables of NNDC website [8].

Nucleus	Cascade	Mixing ratio (δ)
104-Pd	1.342 -- > 0.556 --- >0	>10
152-Sm	0.811 --- > 0.121 ---> 0	>3 ?
168-Er	0.821 --- > 0.07 --- > 0	28^{+12}_{-6}
170-Er	0.934--- > 0.79 ---- >0	>16
182-W	1.221 --- > 0.1 --- >0	30^{+6}_{-4}
184-W	0.903 --- > 0.111 --- >0	16 ± 6
184-W	1.286 --- >1.163 ---- >0	13^{+70}_{-6}
194-Pt	0.622--- > 0.329 --- >0	14 ± 2

4. Application Oriented Research

It is sometimes remarked that any process, scattering or reaction, induced by gamma rays can also be induced by almost all other probes with corresponding energies. Here the point is made that the low photon beam intensities (10^{11} photons/sec or less) and smaller interaction cross sections (at most milli barns) suggest that charged particle probes with hadronic interactions may be better options. While this argument may indeed be correct, the LCS beam facilities offer some specific advantages which make them fertile research venues.

The nearly white intensity distribution of laser scatter photons for up to the end point energies open up new applications and new dimensions to the diverse fields which exploit nuclear techniques. While considering the applications research, it is important to remember that the LCS beamlines are one among the several users at a synchrotron facility and require very little investment in terms of infrastructure and operating costs. From my experience, once installed, routine operation of a LCS facility can be handled by a single operator/scientist. Thus, the operational costs of such facilities are laser maintenance and upgrades and very little man power costs.

4.1. Medical Imaging and Radiation Therapy

The LCS beams of MeV energies have potential to overcome some of the deficiencies of Bremsstrahlung beams for therapeutical applications. A main issue with the Bremsstrahlung radiation is that the dose distribution peaks near the surface and the dose is not localized. The LCS beams, though they do not localize the dose to the extent charged particle beams do, the dose-depth distribution shows a vast improvement. A simulation of white spectrum for 15

MeV endpoint energy shows that the dose maximum occurs at 6 cm interior of the body, compared to 3 cm with conventional Bremsstrahlung. Since there is one to one correspondence of the LCS beam photon and the emission angle, one can further affect the dose-depth distribution by the use of narrowly collimated beams. These aspects of significance for medical applications warrant detailed analyses and experimental verifications. Also, in recent years, there is much interest in developing real time diagnostic imaging during the radiation therapeutic treatments. As seen in figure 1, the dominant interaction mechanism of photons of several MeV is by pair production, suggesting a real time positron emission tomography is achievable. If feasible, it has far reaching consequences for assessing the treatment effects in real time and adjusting the irradiation protocols accordingly. Medical physicists [9] indicate that these beams offer three potential advantages:a) improved beam characteristics for treatment delivery, b) imaging the radiation delivery in real time and c) real time dose verification. Clearly, improvements which impact patient treatments are welcome areas of research.

4.2. *Artificial Transmutation*

As discussed in section 2, photons of 15 MeV energies excite the giant dipole resonance of almost all intermediate mass and heavy nuclei (mass numbers A~ 60-200), which preferentially de-excite by particle emission. Thus, the combination of LCS beams with the physics features of wide giant dipole resonances, makes them a potential tool of artificial transmutations for diverse purposes. A few uses such as production of medical isotopes, nuclear waste management come to mind.

4.2.1. Production of medical isotopes ^{99}Mo and ^{192}Ir

The isotopes ^{99}Mo ($T_{1/2}$= 66 hours) and ^{192}Ir ($T_{1/2}$= 73.8 days) are among the few most commonly used isotopes for SPECT imaging and brachytherapy. The LCS facilities will prove to be among the least expensive venues for the production of these isotopes through^{100}Mo(γ,n) and ^{193}Ir(γ, n) reactions. For ^{99}Mo, it is note worthy that the radiochemical procedures at LCS will constitute a single processing of separating the ^{99m}Tc from the molybdenum isotopes while the conventional method from fission products entails two steps: a) separation of molybdenum from other numerous fission products and b) preparation of the so called 'Moly-cow' to 'milk' the ^{99m}Tc from its parent ^{99}Mo. The first step of this processing is avoided in LCS arrangement. Also, one can perform these experiments with natural molybdenum without having to procure isotopically enriched sample, since there is no other interfering (γ,n) product radioactivity.

4.2.2. Nuclear waste management

In addition to the large volumes and long lived nuclear waste from power plants, we should be concerned about small but finite amounts of radioactive waste from medical facilities, industrial and research establishments. It is not widely recognized that artificial transmutation which are the principal causes of nuclear waste can serve to reduce nuclear waste. For example, the most commonly encountered ^{137}Cs ($T_{1/2}$= 30 years) and ^{60}Co ($T_{1/2}$=5.3 years) can be converted respectively to ^{136}Cs ($T_{1/2}$=13 days) and stable ^{59}Co isotopes by (γ,n) reaction. At facilities such as LCS, the photon beam dumps can be used to irradiate the waste materials for long enough times without any interference for other research activities. Needless to say, the beam dumps must be carefully designed to this end.

4.3. Detection of Contraband Materials

There has been a lot of interest and quite some work is being carried out in nuclear forensics for detecting contraband materials of concern due to illicit trafficking or security risks. In comparison to the X-ray imaging or resonance fluorescence with X-rays for elemental analysis, gamma rays offer distinct advantages despite the fact that the interaction cross sections are smaller. A few strong points of gamma rays are listed below:

- Gamma rays are deep-penetrating and thus probe materials in the deep interiors of containers, not accessible to X-ray probes
- They are much less attenuated by shielding and thus one can probe the interiors of thick walled containers
- The photon- nuclear interactions are isotope-specific and thus we can tell apart hazardous isotopes from the safe ones.
- MeV Gamma rays can cause transmutations, rendering off-line analysis besides the real time data processing possible

The LCS facilities are unlikely to be useful in field testing. However, they have a great potential to establish the scientific and technical principles of such detection systems. There has been some research carried out at the FEL facility of Duke university [10], which shows the promise of photo fission, and resonance fluorescence for these purposes.

5. Positrons and Neutrons as Ternary Beams

In addition to the use of photon beams for material studies and scientific use, the LCS beams can be employed to generate ternary beams, particularly of positrons and quasi-mono energetic fast neutrons.

5.1. *Positron Beams*

As seen above, the predominant interaction of MeV photon with matter is by pair production. It is then possible to generate copious amounts of positrons with a high Z material as converter target. There is an on-going interest in the high resolution micro beams of positrons for surface physics and chemistry studies. More recently, nuclear reactors have embarked in production of positron beams by high energy gamma rays from (n,γ) reactions of reactor neutron interactions in an absorption target. However, the directionality of LCS photons andthat of subsequent positron promises superior quality positron beams.

5.2. *Fast Neutron Beams*

Good quality fast neutron beams for fundamental research and applications are not easily produced. A common technique is a spallation neutron source, which are generated from high energy, high current proton accelerators. For these systems, fast neutrons are still a small component among a huge fluxes of slow neutrons. The LCS gamma rays with endpoint energies on the peak of giant dipole resonances have the potential to produce quasi-mono energetic neutrons with the capability to easily tune the peak energies by a change of production target. Figure 2 shows a calculated neutron energy spectrum of a 15 MeV white spectrum of LCS beam incident on a thin gold target. Clearly seen is a peak at about 6 MeV. It must be pointed that the energy spread can be reduced by collimating the photon beams to allow a small energy spread of the incident photon beam which, in turn, renders the neutron beam energies better defined.

Also, it is simply a matter of changing the production target to change the outgoing neutron beam energies. It is because the neutron separation energies are specific to each isotope, which determine the energies of product neutrons. The table 2 lists a set of production targets, the neutron thresholds and the peak energies for 15 MeV LCS photons. The target materials are easily procured. Also, photon interactions in the target do not have the problem of heat dissipation, which is a huge concern in the spallation neutron sources of high energy, high intensity proton beams.

Table 2. A few select neutron production targets, corresponding thresholds for the (γ,n) reactions and the peak energies of neutrons for LCS beams of E_{max}= 15 MeV.

Production Target	Threshold (MeV)	Neutron Energy(MeV)
197-Au	8	6
120-Sn	9.1	5
63-Cu	10.9	3
90-Zr	12	2
27-Al	13	1

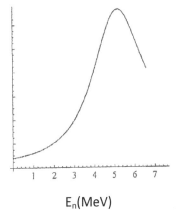

E_n(MeV)

Fig. 2. Neutron Intensity distribution (vertical axis) plotted against neutron energies (E_n) with a white spectrum of LCS gamma rays of 15 MeV endpoint energy incident on a gold target.

6. Summary and Conclusions

During the last 50 years since the first proposal, the Laser Compton scatter photon beams at the synchrotron light sources have matured to become useful tools for nuclear science and technology research. Still, the MeV photon beam facilities are not numerous. At present, the main activities are at the HIγS facility at Duke university (USA) [11] and New Subaru at LASTI, Hyogo, Japan [12]. There are opportunities to build them at the facilities in Canada (Canadian Light Source of 2.9 GeV), India (Indus-2 of 2.5 GeV), and Korea (Pohang light source of 3.0 GeV).

It is pointed out that distinct intensity distribution of these beams offers a unique venue for research and development using the photon beams, which can become competitive with hadrons and neutrons. Several nuclear physics, nuclear astrophysics, security, health etc., related research activities can benefit from studies at these facilities. Also, nuclear research at the light sources will add an extra dimension at these multi disciplinary research facilities.

Acknowledgments

This work was supported in part by a Discovery grant from Natural Sciences and Engineering Council of Canada.

References

1. E. Feenberg and H. Primakoff, Interaction of Cosmic Ray Primaries with Sunlight and Starlight, Physical Review 73 (1948) 449
2. R.E. Milburn, Electron Scattering by an Intense Polarized Photon Field, Physical Review Letters 10 (1963) 75
3. F. R. Arutyunyan and V. A. Tumanian, Quasi-monochromatic and Polarized high-energy gamma rays, Soviet Physics, Uspekhi 7(1964) 339.
4. A. D'Angelo, O. Bartalini, V. Bellini et al., Generation of Compton backscattering γ-ray beams, Nuclear Instruments and Methods, A 455 (2000) 1.
5. Figure adopted from J. Beringer et al (Particle Data Group), Physical Review D86 (2012) 010001
6. T. Shima et al, these proceedings
7. H. Ejiri, these proceedings
8. Nuclear Data Website http://www.nndc.bnl.gov
9. Don Robinson, University of Alberta, and James Robar, Dalhousie University, private communications
10. C. Howell, these proceedings
11. T. Scott Carman, V. Litveninko, and J. Madey et al, The TUNL-FELL inverse Compton g-source as a nuclear physics facility, Nuclear Instruments and Methods in Physics Research, A 378 (2004)1.
12. K. Aoki, K. Hosono, and T. Hadame et al, High-energy photon beam production with laser-Compton backscattering, Nuclear Instruments and Methods in Physics Research A 516 (2004)228.

IRIDE:
Interdisciplinary Research Infrastructure Based on Dual Electron Linacs and Lasers

M. Ferrario[*]

INFN-LNF,

Via E. Fermi 40, 00044 Frascati (Roma), Italy

Massimo.Ferrario@lnf.infn.it

This paper describes the scientific aims and potentials as well as the preliminary technical design of IRIDE, an innovative tool for multi-disciplinary investigations in a wide field of scientific, technological and industrial applications. IRIDE will be a high intensity "particles factory", based on a combination of high duty cycle radio-frequency superconducting electron linacs and of high energy lasers. Conceived to provide unique research possibilities for particle physics, for condensed matter physics, chemistry and material science, for structural biology and industrial applications, IRIDE will open completely new research possibilities and advance our knowledge in many branches of science and technology. IRIDE is also supposed to be realized in subsequent stages of development depending on the assigned priorities.

Keywords: SC linac; fel; particle physics; neutron source; compton source; advanced accelerators concepts.

1. The IRIDE concept: technological breakthroughs as a basis for new research in fundamental and applied science

The proposed IRIDE infrastructure will enable new, very promising synergies between fundamental-physics-oriented research and high-social-impact applications. Conceived as an innovative and evolutionary tool for multi-disciplinary investigations in a wide field of scientific, technological and industrial applications, it will be a high intensity "particles factory", based on a combination of a high duty cycle radio-frequency superconducting electron linac and of high energy lasers. It will be able to produce a high flux of electrons, photons (from infrared to γ-rays), neutrons, protons and eventually positrons and muons, that will be available for a wide national and international scientific

---

[*]On behalf of the IRIDE design study group.

community interested to take profit of one of the most worldwide advanced particle and radiation sources.

We can foresee a large number of possible activities, among them:

- Science with IV generation light sources (IR-X FEL) ;
- Nuclear photonics with Compton back-scattered γ-rays;
- Fundamental physics with low energy linear colliders;
- Advanced neutron source by photo-production;
- Science with THz radiation sources;
- Physics with high power/intensity lasers;
- R&D on advanced accelerator concepts ;
- International Linear Collider technology implementation;
- Detector development for X-ray FEL and Linear Colliders;
- R&D in accelerator physics and industrial spin off.

The main feature of a superconducting (SC) linac relevant for our facility is the possibility to operate the machine in continuous (CW) or quasi-continuous wave (qCW) mode with high average beam power (< 1 MW) and high average current (< 500 μA). The CW or qCW choice, combined with a proper bunch distribution scheme, offers the most versatile solution to provide bunches to a number of different experiments, as could be envisaged in a multipurpose facility. In addition Europe is in a strategic position in the SC RF technology, mainly due to the strong contribution of European countries to the TESLA Collaboration [2]. In particular INFN strongly participated since the early design stages through the final engineering, shares the know-how with Italian industries (see Figure 1) and has the recognized intellectual property of several main components (one of which is the cryo-module concept and its evolution [3]).

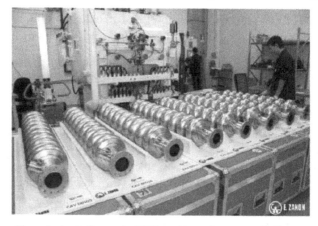

Figure 1: A set of pre-series XFEL cavities at the company E. Zanon.

The realization of such a large facility will allow INFN to consolidate a strong scientific, technological and industrial role in a competing international context both to deploy a national multipurpose facility along the scientific applications discussed in the following sections, and to prepare a strong role for the contribution to possible future large international high energy physics projects such as the International Linear Collider [4].

2. IRIDE layout: staging and upgrade potentials

The backbone of the IRIDE facility is a double superconducting high duty cycle electron linear accelerator, with the required 15 kW at 2 K cryogenic plant, based on the L-band standing wave RF (1.3 GHz) cavities developed by the TESLA collaboration, which currently drive the FLASH FEL facility in DESY and which, with minimal improvements of the cryo-module cooling system, could be upgraded to CW or qCW operation, see Table 1. Both pulsed and CW options rely on existing technology, available on the market by several vendors. Pulsed klystrons at 1.7 MW peak delivering 1.5 ms pulses at 100 Hz repetition rate are for instance available from Thales (TH2104D). Solid State Amplifiers delivering 5-10 kW CW power are available from several Italian or European vendors. XFEL/LC couplers exceed the IRIDE parameter requirements in term of both peak and average power.

Table 1. Possible SC linac parameters (for each linac).

	Pulsed	qCW	CW
Energy [GeV]	2	2	1.5
I (within pulse) [mA]	2.5	0.26	
I (average) [mA]	0.17	0.16	0.35
RF pulse duration [ms]	1.5	1000	CW
RF duty cicle [%]	15	60	100
E_{acc}[MV/m]	20	20	15
Q_o x 10^{10}	2	2	2
Q_{ext} x 10^6	4	40	40
Number of cavities	96	96	96
Number of modules	12	12	12
Beam average power [kW]	334	309	525

The second core device of the facility is the high average power cryogenically cooled Yb:YAG Laser system operating in a chirped pulse amplification architecture followed by a frequency conversion stage to achieve 515 nm wavelength. This technology allowed achieving recently 1 J at 100 Hz in the picosecond regime with a bandwidth of 0.1%. An upgraded version of the

SPARC-LAB [5] laser FLAME [6], based upon a Ti:Sa chirped pulse amplification (CPA) laser, will be also considered, leading to a performance characterized by a total energy of 12 J at 10 Hz, with a bandwidth sufficient to achieve < 20 fs and a contrast as high as 10^{12}, suitable to drive advanced plasma wake field acceleration schemes.

By using standing wave SC accelerating structures, that can accelerate beams in both longitudinal directions, one can see an attractive scheme based on two linacs operating at a maximum energy of 2 GeV each, when working in the collider mode, or used in cascade, as a single longer linac, to boost the electron energy up to 4 GeV for higher energy electron beam applications. In addition when operating in the collider mode both linacs may partially recover the electron kinetic energy of the beam leaving the opposite linac after the interaction, thus increasing the overall efficiency of the system and simplifying the beam dump design.

The IRIDE design options have fundamental implications on the electron injector and on the electron gun. Indeed, the injector accelerating sections must be of the superconducting type, as the rest of the electron linac, while the gun can be in both cases a sub-harmonic, low frequency RF gun, operating at room temperature. The results recently obtained at Berkeley with the CW APEX gun [7], in the framework of NGLS [8], are consistent with the CW option and can also be extrapolated with the required brightness in case the high charge bunch, high repetition rate option is chosen. The possibility to produce polarized electrons with the APEX gun is under investigation.

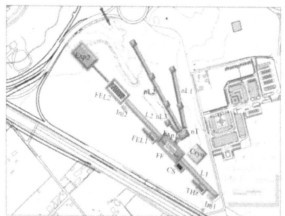

Figure 2: Schematic layout of the IRIDE accelerators and radiation sources complex. Some relevant components are indicated: the cryogenic plant hall (Cryo), the electron injectors (Inj1, Inj2), the SC linacs (L1, L2), the FEL devices (THz, FEL1, FEL2) and the user experimental hall (Exp1,Exp2), the laser system for Compton source (CS), the Final Focus beam line (FF) in the experimental hall Exp1 for colliders and nuclear photonics, the neutron source target (nT) with experimental beam lines (nL1, nL2, nL3).

As indicated in Fig. 2, the first 2 GeV linac system (L1) can drive FEL, Neutron and THz radiation sources, electron-on-target physics experiments and, in combination with the high energy laser, a 10-60 MeV γ-ray Compton source is also possible. With also the second 2 GeV linac installed (L2) one can envisage a low energy linear collider scheme for electron-electron, electron-photon, photon-photon and eventually electron-positron scattering studies. The combination of the two linacs, boosting the electrons up to 4 GeV, could also drive a short wavelength FEL user facility (Exp2).

The *Neutron source* (nT) requires a medium energy electron beam such as the one extracted from L1 driven to impinge on bulk of cooled high-Z target (nT), where it loses energy mainly by bremsstrahlung, producing an electromagnetic shower cascade. The photons of the shower can excite the nuclei of the target with which they interact and these excited nuclei go back into the fundamental state by emitting one or more nucleons. At the state of the art of the project, we have mainly focused on the Tungsten as possible choice for the target of IRIDE: the estimated rate emissions of neutrons (up 10^{15} n/s) and other secondary particles, that are described in [1], have been obtained for a Tungsten cylindrical cooled target with 7 cm diameter and 6 cm height. A 3 m thick Iron shielding is also foreseen. The beam-lines of interest are of three types: short beam-lines for Chip Irradiation and Imaging (nL3), long beam-lines for applications requiring time of flight measurements, like Bragg Edge Transmission, Diffraction and Nuclear Resonance Capture Analysis (nL1), and even longer beam-lines (~200 m) needed for neutron oscillation studies (nL2). Each beam line needs to be equipped with shielding, diagnostics and detectors.

The main components of the *Compton Source* (CS) are [9]:

- the high brightness electron Linac (L1) capable to deliver multi-bunch trains, i.e. working at 100 Hz rep rate with at least 50 electron bunches distributed over the RF pulse duration (from 0.5 to 1 μs), carrying a fraction of a nC bunch charge at very low rms normalized emittances (< 1 mm-mrad) and energy spreads (< 0.1%);
- a high energy, high quality, high repetition rate laser system, delivering pulses carrying at least 1 J of energy (in the fundamental), psec pulse duration, 100 Hz repetition rate, high quality ($M^2 < 1.2$), such to be focused down to typical spot sizes of 10 μm at collision with the electron bunch;
- a laser recirculator consisting of a two parabolic confocal mirror set, capable to recirculate the laser pulse a number of times equal to the electron bunches within the train (< 50), by focusing it down to the collision point, recollimating and reflecting it back to the other mirror which in turns refocuses it down back to the interaction.

The expected performances for the γ-ray beam delivered are: tunability between 1 and 60 MeV, bandwidth smaller than 0.3%, full control of polarization (linear, larger than 99.8%), spectral density larger than 10^4 photons/s·eV, and peak brilliance larger than 10^{22} (photons/s-mm²-mrad²-0.1%).

Several *FEL source* configurations are possible at IRIDE ranging from IR to X ray wavelength radiation, as discussed in Ref.[1]. In particular, a seeded configuration is possible at the exit of L1, by using an externally injected laser signal. In this case the maximum operating energy is fixed by the source exploited as seeding and by the undulator parameters. If we consider the 27-th harmonics of the Ti-Sa (26.9 nm), the beam energy is constrained below 1 GeV to drive on resonance an undulator with 2.7 cm period. The FEL tunability could range from 27 nm to 1.65 nm (FEL1). Possible options with shorter undulator period are under investigation including the RF undulator option. User beam lines can be accommodated in the first experimental hall (Exp1).

At the highest energy end of the linac (FEL2) a combination of Oscillator, SASE and Seeded operational mode offer an attractive and unique possibility. As shown in Figure 3 an oscillator operating in the VUV region is used to produce bunching in the e-beam train, which is successively injected into the downstream sections of the undulator chain tuned at higher harmonics of the oscillator [10]. The rather narrow bandwidth of operation of the cavity provides a constraint for the energy of the e-beam at an energy around 2.28 GeV. In this configuration a significant amount of third and fifth harmonics allows its use in the successive section to get pre-bunched SASE operation at 4.5 and 2.7 nm. Removing the cavity mirrors and operating at full linacs energy (up to 4 GeV, see Table 2) we get an output wavelength of 0.6 nm, which can be extended to 0.2 nm provided that a segmented undulator is installed and the last sections are replaced by a super-conducting undulator with undulator parameter $K_u = 1$ and period $\lambda_u = 1$ cm.

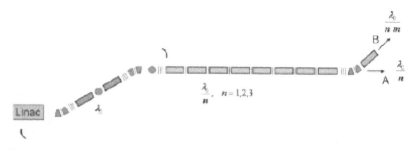

Figure 3: Undulator chain at the L2 exit, the first component is an oscillator acting also as a micro-buncher driving the downstream SASE FEL.

The possibility to operate the IRIDE SASE FEL in a qCW mode with moderate repetition rate (<1 MHz) will certainly simplify the required x-ray detector performances. User beam lines can be accommodated in the second large experimental hall (Exp2).

Table 2. FEL performances at 4 GeV electron beam energy

	Fundamental	3rd Harmonic
Resonant Wavelength λ_r [nm / keV]	0.6 / 2	0.6 / 2
Peak flux [photons/s-0.1% BW]	1.2×10^{25}	5.9×10^{22}
Peak brilliance	1.9×10^{31}	1.8×10^{29}
Photons/bunch	2.1×10^{12}	1.1×10^{10}

Experimental hall 1 (Exp1) will host also the particle detectors for the linear collider options, certainly the most challenging components of the entire project. The *electron-electron collider* option will be essentially based on the final focus (FF) system already operating at ATF2 [11], based on the recently proposed compact final focus optics with local chromaticity correction [12] and where 100 nm electron beam spot sizes have been already achieved. The occurrence of parasitic collisions in the collider mode will be avoided with a crossing angle. The feasibility of the *electron-positron collider* is strongly dependent on our capability to produce low emittance positrons [13], [14]. The IRIDE R&D program comprises the development of a positron source based on direct conversion of a 60 MeV γ-ray beam in a solid target. The positron source design goals are reported in Table 3.

Table 3. Comparison between conventional and expected Compton driven positron source performances

	Conventional	ICS Driven
rms source size [μm]	400	50
Target thickness [X_o]	6	0.4
rms transverse momentum [MeV]	5	1
rms norm. emittance [m-rad]	0.001	50×10^{-6}
Positron yield	1	0.4

The *gamma-electron* and *gamma-gamma collider* options will require a careful design and development of the interaction region (see Fig. 4 [15]). In a photon collider in fact two high energy electron beams after the final focus system travel towards the interaction point (IP) and at a distance of about 1- 5 mm from the IP collide with a focused laser beam [16]. After scattering, the photons follow their direction to the interaction point (IP) where they collide with a similar opposite beam of high energy photons or electrons. Such a new collider configuration has never been realized so far and is the subject of many design studies around the world. A dedicate design for the IRIDE facility is under way.

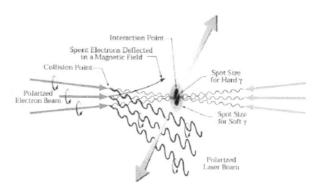

Figure 4: Scheme of principle of a photon collider. High-energy electrons scatter on laser photons and produce high-energy photon beam which collides with a similar photon or electron beam at the interaction point IP [15].

Advanced accelerators techniques could be also investigated in the large experimental hall (Exp1). The success of the advanced accelerator activity as a vigorous and intense R&D program focused on the enabling technologies of plasma accelerators, Compton converters, gamma beam focusing, polarized positron source, superconducting RF gun and the associated advanced diagnostics instrumentation, could allow envisaging a convenient energy upgrade of the facility to tens GeV level in a higher energy range of scientific applications.

The IRIDE facility could be hosted in the 30 hectares area on the University of Rome Tor Vergata campus site a few km southeast of the city of Rome as shown in Fig. 2. The total linear extension will be approximately 700 m. The interested area is just alongside of the CNR territory and it is approximately a couple of km away from the ENEA and INFN sites in Frascati, the major Italian research institutes that have already strongly contributed to IRIDE scientific case.

IRIDE is also supposed to be realized in subsequent stages of development depending on the assigned priorities and available fundings. An initial stage could include the first linac only, up to a maximum energy of 2 GeV, which could drive the realization within 5 years of the neutron source, the long wavelength FEL, the Compton source and the fixed target nuclear physic studies. Provided that the average beam current is kept constant, all the previous applications could be run concurrently.

An overview of the IRIDE scientific case is reported with more details and with a complete reference list in Ref. [1].

References

1. M. Ferrario *et al*, Nucl. Instr. and Meth. A 740 (2014) 138–146, see also D. Alesini *et al.*, IRIDE White Book, arXiv:1307.7967 [physics.ins-det].
2. B. Aune *et al.*, Superconducting TESLA cavities, Phys. Rev. ST Accel. Beams 3, 092001 (2000).
3. C. Pagani *et al.*, Construction, Commissioning and 3genic Performances of the First TESLA Test Facility (TTF) Cryomodule, Advances in Cryogenic Engineering, Vol 43, Edited by P. Kittel, Plenum Press, New York, 1998, pp. 87.
4. T. Behnke *et al.*, The International Linear Collider Technical Design Report - Volume 1: Executive Summary, arXiv:1306.6327 [physics.acc-ph].
5. M. Ferrario *et al.*, Nucl. Instr. and Meth. B 309, 183 (2013).
6. L. Gizzi *et al.*, Nucl. Instr. and Meth. B 309, 202 (2013).
7. F. Sannibale *et al.*, Phys. Rev. ST Accel. Beams 15, 103501 (2012).
8. J. M. Byrd *et al.*, Design Concepts for the NGLS linac, Proceedings of IPAC13, Shanghai, China.
9. L. Serafini *et al.*, ELI-NP GBS Technical Design Report (to be published).
10. G. Dattoli and P. L. Ottaviani, J. Appl. Phys. 86, 5331 (1999).
11. Sha Bai *et al.*, First beam waist measurements in the final focus beam line at the KEK Accelerator Test Facility, Phys. Rev. ST Accel. Beams 13, 092804 (2010).
12. P. Raimondi and A. Seryi, Phys. Rev. Lett. 86, 3779 (2001).
13. K. Floettmann, Positron source options for linear colliders, Proceedings of EPAC 2004, Lucerne, Switzerland.
14. E. Sabia et al., Free Electron Laser and Positronium Stimulated Annihilation, Proceedings of FEL 2010, Malmo, Sweden.
15. K.-J. Kim and A. Sessler, Gamma-Gamma Colliders, Beam Line, Spring/Summer 1996.
16. V. Telnov *et al.*, The Photon Collider at Tesla, Int. J. Mod. Phys. A 19, No. 30, 5097 (2004).

Poster Presentations

A Study of the Nuclear Resonance Fluorescence Reaction Yield Dependence on the Target Thickness of ^{208}PB

Hani Negm[*], Izuru Daito, Heishun Zen, Toshiteru Kii, Kai Masuda,
Toshitada Hori and Hideaki Ohgaki

Institute of Advanced Energy, Kyoto University, Uji, Kyoto 611-0011, Japan

Ryoichi Hajima, Toshiyuki Shizuma, Takehito Hayakawa
and Nobuhiro Kikuzawa

*Quantum Beam Science Directorate, Japan Atomic Energy Agency,
Tokai, Ibaraki, 319-1195, Japan.*

Hiroyuki Toyokawa

*Research Institute of Instrumentation and Frontier, National Institute of Advanced
Industrial Science and Technology, Tsukuba, Ibaraki, 305-8561, Japan.*

We have been developing an active, non-destructive detection system based on nuclear resonance fluorescence (NRF) for inspecting special nuclear materials (SNMs) such as ^{235}U in a container at a seaport. The study of the NRF yield dependence on the target thickness of SNMs is required to evaluate the performance of the inspection system. To this end, an NRF experiment has been performed using a laser Compton backscattering γ-ray beam line at New SUBARU in ^{208}Pb. Cylindricalshaped natural lead targets with a 0.5 cm radius and varying thicknesses of 1.0, 1.44, and 3.05 cm were irradiated at a resonance energy of 7.332 MeV. The NRF yield was detected using two HPG edetectors with relative efficiencies of 120% and 100% positioned at scattering angles of 90°and 130°, respectively, relative to the incident γ-ray beam. As a result, the NRF yield exhibited a saturation behavior for the thick lead target. An analytic treatment and Monte Carlo simulation using GEANT4 was performed to interpret the reaction yield (RY) of the NRF interaction. The simulation result is in good agreement with the experimental data for the target thickness dependence. The analytic treatment, the NRF RY model, is also in reasonable agreement.

---

[*] Corresponding Author: Hani Negm (haninegm@iae.kyoto-u.ac.jp)

1. Introduction

Threats of smuggling SNMs or nuclear weapons hidden in a cargo container have motivated a number of laboratories worldwide to develop inspection systems for hidden SNMs. The utilization of bremsstrahlung γ-rays [1-2] and neutrons[3-4] as probes to measure the isotopic composition of heavily shielded materials has been studied. Recently, Pruet et al. proposed a novel, nondestructive detection method for ^{235}U concealed in a cargo transporter by using laser Compton scattering (LCS) γ-rays [5]. The LCS γ-ray has characteristics of energy-tunable, monochromatic γ-rays by scattering laser pulses from relativistic electrons [6-8]. This process is known as Thomson or inverse Compton scattering and produces γ-ray pulses that have beam-like characteristics and exhibit exceptional brightness. We also have proposed an inspection system, which is based on a technique that is a hybrid of neutron and γ-ray interactions to detect hidden SNMs in a cargo container [9]. It consists of a quick pre-screening using D-D neutrons generated by an inertial electrostatic confinement (IEC) device [10-11] and a post-screening based on nuclear resonance fluorescence (NRF) using quasi-monochromatic γ-rays generated by LCS.

NRF provides a unique approach to identifying material compositions on an isotopic basis. NRF is a phenomenon in which a photon of an appropriate energy excites a nucleus to a higher state, which subsequently decays to the ground state or a lower-lying state by emitting a γ-ray with an energy equal to the energy difference between the resonance state and the lower state [12-13].

In our proposal of the inspection system for hidden nuclear materials, we assumed that a certain amount of SNMs with a high purity would be used. Thus, in this paper, we will study the dependence of the NRF reaction yield on the target thickness, which is on the order of several centimeters. For this purpose an NRF experiment, which uses an LCS γ-ray beam and HPGe detectors, on the heavy isotope ^{208}Pb was conducted. In addition, a model for NRF reaction yield (NRF RY model) is introduced to interpret the experimental data. In addition, a Monte Carlo simulation using GEANT4 [14-15] was performed to simulate the experimental geometry for ^{208}Pb. Section 2 presents the results of the NRF experiment on ^{208}Pb at the 7.332 MeV energy level for various thicknesses. In Section 3, the analytic treatment of the NRF RY will be introduced to describe the experimental results. Sections 4 present the simulation results of the Monte Carlo simulation toolkit GEANT4 for a simulation. The simulation results are compared to the experimental data for ^{208}Pb in Section 4. Both the experimental and simulation results are compared to the NRF RY model for ^{208}Pb.

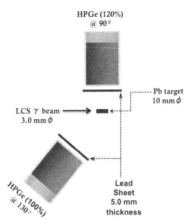

Figure 1. Schematic of the experimental setup for the NRF measurements of the ²⁰⁸Pb target.

2. Experimental Setup

NRF measurements of ²⁰⁸Pb were performed at the LCS γ-ray beam line (BL1) at New SUBARU [16]. The γ-ray energy was ~7.4 MeV, and the beam had an energy spread in full width at half maximum (FWHM) of 10%. The γ-ray beam was collimated by a lead collimator with a 0.3 cm hole, which defines the beam size as 0.3 cm. Three cylindrical natural lead targets with thicknesses of 1.0, 1.44, and 3.05 cm and a radius of 0.5cm were used. The NRF γ-rays were measured using two HPGe detectors with relative efficiencies of 120% and 100% situated at 90° and 130°, respectively, with respect to the incident beam axis and focused on the geometric center of the Pb target as illustrated in Figure 1. A 0.5 cm sheet of lead was placed in front of each detector to attenuate the low energy region of the γ-rays scattered by the target to reduce the count rate of the HPGe detector.

2.1. Experimental Results

Typical spectra of the NRF measurements of ²⁰⁸Pb at a resonance energy of 7.332 MeV are shown in Figure 2. The upper spectrum is taken from data from the HPGe detector at 90°for the 1.44 cm Pb target, and the lower spectrum is that from the 130° HPGe detector. The two NRF resonance levels were observed at 7.083 and 7.063 MeV in addition to the 7.332 MeV level. The remaining peaks in the spectrum are the single and double escape peaks of the NRF γ-rays of the observed states. In this study, we focused on the 7.332 MeV energy level. The normalized count rate of allruns is illustrated in Figure 3.

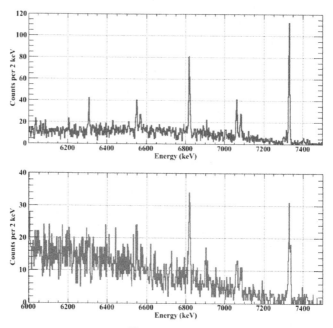

Figure 2. Measured γ-ray energy spectra of ^{208}Pb for a 1.44 cm-thick target. The upper histogram is for the 90° scattering angle, and the lower is for the 130° scattering angle.

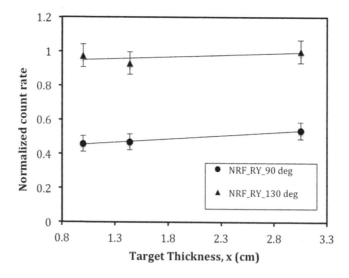

Figure 3. The relation between the NRF count rate (reaction yield) and the target thickness for ^{208}Pb ata resonance energy of 7.332 MeV for three thickness x = 1.0, 1.44, and 3.05 cm.

The normalization was performed with respect to the measurement live-time and the total flux of the incident γ-ray beam which was monitored by 8' NaI(Tl) detector placed downstream of the beamline. The detection efficiency of the detectors was calibrated using ^{60}Co and subsequently extended to high energies using GEANT4. As shown in Figure 3, it is clear that the NRF reaction yields are almost constant in this thickness range.

3. NRF Reaction Yield Model

The NRF reaction yield R can be expressed by [18]

$$\frac{d^2 R}{dxdE} = \Phi_i \cdot e^{-\mu_a x} \cdot N \cdot \sigma \cdot W(\theta) \cdot e^{-\mu_s r_s} \cdot \varepsilon(E). \frac{\Omega}{4\pi} \qquad (1)$$

Where Φ_i is the incident γ-ray beam flux, which attenuates in the target material by $exp(-\mu_a x)$, and where μ_a is the absorption coefficient for the incident beam. In the case of nuclear resonance, μ_a involves the atomic attenuation and the nuclear attenuation by nuclear resonance absorption. N is the atomic density of the isotope of interest in the target, and σ is the NRF cross-section of a nuclear state, which is given by the Breit-Wigner distribution [12]. $W(\theta)$ is the angular distribution of the scattered NRF γ-rays [18], which attenuate by $exp(-\mu_s r_s)$, where μ_s represents the scattering attenuation coefficient for the scattered NRF γ-rays and r_s is the scattering path length. The scattering NRF γ-rays can't be reabsorbed by the resonance because some of the energy of the emitted NRF γ-rays goes to the recoiling nucleus. $\varepsilon(E)$ is the detection efficiency of the γ-ray detector, and $\Omega/4\pi$ is the fraction of the solid angle subtended by the detector.

Figure 4 illustrates the slab geometry of the NRF interaction. We assumed a parallel incident γ-ray beam on the surface of a target with a thickness L, and the beam size and the target thickness are less than the distance between the detector and the target. The attenuation of the γ-rays consists of the absorption and scattering terms. Therefore, the integration of Eq. (1) for the absorption is over the target thickness and the scattering path length that is determined by the scattering angle θ between the incident beam and scattering angle of the detector. Given $r_s = x / \cos(\theta)$ and $\theta \neq 90°$, we have

$$\int_0^L e^{-\mu_a x} \cdot e^{-\mu_s r_s} \cdot dx \Rightarrow \int_0^L e^{-(\mu_a + \frac{\mu_s}{\cos\theta})x} \cdot dx \Rightarrow \frac{1 - e^{-\mu L}}{\mu} \qquad (2)$$

where $\mu(E_r) = \mu_a + \frac{\mu_s}{\cos\theta}$, $\mu_a(E_r) = \mu_{atomic} + \mu_{NRF}$, $\mu_s(E_r) = \mu_{atomic}$, and $\quad \mu_{NRF} = \dfrac{N_a}{A} \sigma_{NRF}$

where A is atomic mass of isotope of interest and N_a is the Avogadro's number. For $\theta = 90°$, $r_s \rightarrow b$, where b is the distance from the center of the target to its boundary; e.g., in the case of cylindrical or spherical shapes, b represents the radius of the shape, and in the case of a cube, b represents half of the length of a side of the cube. In this case, the attenuation term can be expressed as

$$\int_0^L e^{-\mu_a x} \cdot e^{-\mu_s r_s} \cdot dx \Rightarrow e^{-\mu_s b} \int_0^L e^{-\mu_a x} \cdot dx \Rightarrow \frac{1 - e^{-\mu_a L}}{\mu_a} e^{-\mu_s b} \tag{3}$$

Because the energy of interest is very narrow because of the level width of the NRF, the integration over E can be simplified as

$$\int \sigma(E) dE = \sigma_{NRF}(E_r) \Gamma_D \tag{4}$$

where Γ_D is the Doppler width of a broadened resonance [12]. Substituting Eq. (2) and Eq. (4) into Eq. (1), we obtain the NRF reaction yield R

$$R = \underbrace{\left[\Phi_i(E_r) \cdot N \cdot \sigma_{NRF} \cdot \Gamma_D \cdot W(\theta) \right]}_{\text{Non Attenuated NRF RY}} \cdot \underbrace{\left[\varepsilon(E_r). \frac{\Omega}{4\pi} \right]}_{\text{Efficiency Term}} \cdot \underbrace{\left[\frac{1 - e^{-\mu L}}{\mu} \right]}_{\text{Attenuated Factor}} \tag{5}$$

This equation can be divided into three terms. The first term is the non-attenuated reaction yield. The second term is the detection efficiency and the solid angle subtended by the detector. The last term is the attenuation factor for the NRF reaction yield. The attenuation length (X_0) or the half value layer ($X_{1/2}$) of NRF RY curve can be easily evaluated by the total attenuation coefficient, $\mu(E_r)$, where $X_0 = \ln(2.72)/\mu(E_r)$, and $X_{1/2} = \ln(2)/\mu(E_r)$.

In the case of a thin target, the attenuation term can approximated as

$$\frac{1 - e^{-\mu L}}{\mu} \approx \frac{1 - (1 - \mu L)}{\mu} = L \tag{6}$$

Therefore, the reaction yield for a thin target can be written as

$$R = \Phi_i(E_r) \cdot N \cdot \sigma_{NRF} \cdot \Gamma_D \cdot W(\theta) \cdot \varepsilon(E_r). \frac{\Omega}{4\pi} \cdot L \tag{7}$$

This equation shows the linear trend for the NRF reaction yield for the thin targets that cause small attenuations for the incident beam and scattered NRF γ-rays. In the next section, we will introduce the experimental and MC simulation data of the NRF reaction yield, which can be described by the reaction yield model.

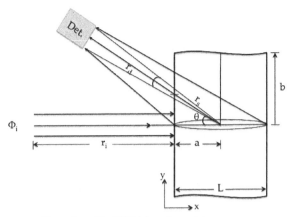

Figure 4. Slab geometry illustration of the NRF interaction point.

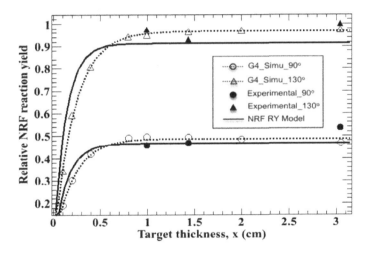

Figure 5. The relative NRF reaction yield of ^{208}Pb (7.332 MeV) as a function of the target thickness, where two sets of the data are shown. The experimental (for thicknesses of x = 1.0, 1.44, and 3.05 cm) and simulated (for target thicknesses of x = 0.1, 0.2, 0.4, 0.8, 1.0, 1.44, 2.0, and 3.05 cm) results at the two scattering angles 90° and 130°. In addition, the NRF reaction yield model was verified for both angles.

4. Monte Carlo Simulation Using GEANT4

The MC simulation toolkit GEANT4 was developed to simulate NRF interactions [19]. The angular correlation function of the scattered NRF γ-rays and the recoiling energy of a nucleus are taken into account by us. The simulation was applied to the same geometry as in the experimental setup described in Section 2. The energy spread (at FWHM) was assumed to be 36.66 keV for the 7.332 MeV incident beam to reduce the calculation time.

Figure 5showsthe simulation results of the scattered NRF γ-rays at the two angles 90° and 130°for the different target thicknesses as well as the experimental results. The NRF reaction yield model applied for these experimental conditions is shown in Figure 5 and is denoted by a solid line. The simulation results are in good agreement with the experimental data. Here it should be noted that the energy shift by the recoiling effect of NRF interaction is important to simulate the NRF yield, because the attenuation of the NRF interaction is the main term of the total attenuation coefficient as seen in previous section.

Figure 5 also shows that the reaction yield model is in reasonable agreement for both the experimental and simulation data. The NRF reaction yield shows a saturation behavior after a certain thickness, which can be expected from the NRF RY model because of the attenuation term. For example, for ^{208}Pb target at resonance energy 7.332 MeV, the absorption coefficient 7.76 cm^{-1} and the scattering coefficient at scattering angle 130° is 0.82 cm^{-1}, therefore the attenuation length for NRF RY is 0.12 cm. On the other hand, the attenuation length from the fitting of the simulation result of NRF RY curve of ^{208}Pb is 0.22 cm. The difference between the NRF RY model and the simulation result could be attributed to the assumption that the solid angle subtended by the detector is independent of the interaction location/volume within the target. It should be noted that the experimental data in thin target region is required for a further discussion.

5. Conclusion

The NRF reaction yield dependence on the target thickness has been investigated. The experimental results are in good agreement with the MC simulation using GEANT4for a thick ^{208}Pb target. The reaction yield model was also deduced analytically and is in good agreement with the experimental results. All the results, i.e., the experimental data and the calculations, show that the NRF yield increases with increasing target thickness until it saturates after a specific thickness which is expressed by the total attenuation of the resonant photons through the target. It is clear that the main term in the total attenuation is the attenuation by the NRF interaction. The attenuation length deduced from GEANT4 simulation is about twice larger than analytical calculation. Thin target experiment is required for precise discussion.

Acknowledgments

This work was supported in part by special coordination funds for promoting science and technology in Japan (grant no. 066).

References

1. W. Bertozzi and R. J. Ledoux, *Nucl. Instrum. Methods Phys. Res.* **B 241**, 820 (2005).
2. W. Bertozzi, S. E. Korbly, R. J. Ledoux, and W. Park, *Nucl. Instrum. Methods Phys. Res., Sect.* **B 261,** 331 (2007).
3. D. A. Close, L. A. Franks, and S. M. Kocimski, *Nucl. Instrum. Methods Phys. Res.* **A 220**, 531 (1984).
4. M. Haruyama, M. Takase, and H. Tobita, *J. Nucl. Sci. Technol.* **45**, 432 (2008).
5. J. Pruet, D. P. McNabb, C. A. Hagmann, F. V. Hartemann, and C. P. J. Barty, *J. Appl. Phys.* **99**, 123102 (2006).
6. H. Ohgaki, H. Toyokawa, K. Kudo, N. Takeda, and T. Yamazaki: *Nucl. Instrum. Methods Phys. Res.* **A 455**, 54 (2000).
7. F. V. Hartemann, W. J. Brown, D. J. Gibson, S. G. Anderson, A. M. Tremaine, P. T. Springer, A. J. Wootton, E. P. Hartouni, and C. P. Barty: *Phys. Rev. ST Accel. Beam* **8**, 100702 (2005).
8. K. Kawase, M. Kando, T. Hayakawa, I. Daito, S. Kondo, T. Homma, T. Kameshima, H. Kotaki, L.-M. Chen, Y. Fukuda, A. Faenov, T. Shizuma, M. Fujiwara, S. V. Bulanov, T. Kimura, and T. Tajima, *Rev. Sci. Instrum.* **79**, 053302 (2008).
9. H. Ohgaki, et al, *J. of the Korean Physical Society.* **59**, 5, 3155-3159 (2011).
10. R. L. Hirsch, *J. Appl. Phys.* **38**, 4522 (1967).
11. G. H. Miley, L. Wu and H. J. Kim, *J. Radioanal. Nucl.Chem.* **263**, 159 (2005).
12. F.R. Metzger, *Prog. in Nuc. Phys.* **7**, 54 (1959).
13. U. Kneissl, H. H. Pitz, and A. Zilges, *Prog. Part. Nucl. Phys.* **37**, 349 (1996).
14. S. Agostinelli, *Nucl. Instr. and Meth.* **A 506**, 250 (2003).
15. J. Allison et al., *IEEE TNS*, **53**, 1 (2006).
16. S. Miyamoto et al., *Radiation Measurements.* **41**, S179 – S185 (2007).
17. Quiter, Brian J., B.A. Ludewig, V.V. Mozin, C. Wilson, S. Korbly, *Nucl. Instrum. Methods Phys. Res., Sect.* **B 269,** 1130–1139 (2011).
18. L.W. Fagg, et al., *Rev. Mod. Phys.* **2**, 711-758 (1959).
19. T. Hayakawa et al., *INMM 52th Annual Meeting Proceeding*, (2011).

Photodisintegration Reactions with Linear Polarized γ-Ray Beam[*]

T. Hayakawa

Japan Atomic Energy Agency, Shirakata-shirane 2-4, Tokai, Ibaraki 319-1195, Japan
National Astronomical Observatory in Japan, Mitaka, Tokyo 181-8588, Japan.
[†]*hayakawa.takehito@jaea.go.jp*

S. Miyamoto, K. Horikawa, S. Amano and T. Mochizuki,
University of Hyogo
Hyogo 678-1205, Japan

K. Imazaki, D. Li and Y. Izawa
Institute for Laser Technology
2-6 Yamada-oka, Suita, Osaka 565-0871, Japan

K. Ogata
RCNP, Osaka University
1-1 Yamadaoka, Suita, Osaka 565-0871, Japan

S. Chiba
Tokyo Institute of Technology
2-12-1 Ookayama, Meguro-ku, Tokyo 152-8550, Japan

Photoreactions are applied for a useful assay tool to detect hidden nuclear materials. If we can employ polarized γ-rays, the $(\vec{\gamma}, n)$ reactions would be used to detect hidden nuclear materials with a higher signal-to-noise ratio. In 1950's, Agodi predicted that the angular distribution of cross sections in (γ, n) reactions with a 100% linearly polarized γ-ray beam for dipole excitations should be anisotropic and universally described by a simple function of $a + b \sin(2\phi)$ at polar angle $\Phi = 90°$. However, there is no experimental $(\vec{\gamma}, n)$ reaction data with linear polarized photons except some light nuclei such as deuteron. We have verified experimentally the fact that this anisotropic angular distribution is manifested on ^{197}Au, ^{127}I, and natural Cu using linearly polarized laser Compton scattering γ-rays at NewSUBARU. We have measured neutron energy using a Time-Of-flight method. We have changed the angle of linear polarized plane of the incident laser. Neutron angular distributions on the three targets can be well reproduced by the formula predicted by Agodi. We have verified the Agodi's prediction over the wide range region for the first time.

Keywords: Photonuclear reactions; laser Compton scattering gamma-rays.

---

[*] This work is supported by in part by Grants-in-Aid for Scientific Research (25610061, 24340060, 22740169, 23654204) of Japan.

1. Introduction

The nuclear resonance fluorescence with a γ-ray beam has been proposed a method to measure fissionable isotopes as plutonium or uranium [1-5]. Another candidate of a nuclear reaction to detect hidden nuclear materials is photodisintegration reactions. This method has an advantage that one can measure neutrons from the photodisintegration reactions with a high signal-to-noise (S/N) ratio by using a time gate because speeds of neutron with energies of several MeV is roughly 1/10 of the light speed. The $(\vec{\gamma}, n)$ reactions have also a chance to detect with high S/N ratio as nuclear resonance fluorescence. In 1950's, Agodi predicted that the angular distribution of cross sections in (γ, n) reactions with a 100% linearly polarizedγ-ray beam for dipole excitation should be anisotropic and universally described by the simple function of $a + b \sin(2\phi)$ at polar angle θ=90 degree [6]. However, there is no experimental data with linear polarized photons except some light nuclei such as deuteron. We have verified experimentally the fact that this anisotropic angular distribution is manifested on ^{197}Au, ^{127}I, and natural Cu using linearly polarized laser Compton scatteringγ-rays at NewSUBARU [7]. We have measured neutron energy using a Time-Of-flight method. We have changed the angle of linear polarized plane of the incident laser. The neutron angular distributions on the three targets can be reduced by the formula predicted by Agodi. We have verified the Agodi's prediction over the wide angularrange for the first time.

2. Experimental Procedure

We have performed a nuclear experiment at NewSUBARU in SPring-8. This facility is managed by the university of Hyogo. Figure 1 shows a schematic view of photonuclear reactions on the nuclide ^{197}Au as an example. Excited states on ^{197}Au are populated by induced γ-rays. Individual populated states decay predominantly to the ground state or an excited state in the neutron deficient isotope ^{196}Au by emitting a neutron. Theγ-ray beam was generated by Compton scattering laser photons with electrons stored in the electron storage ring NewSUBARU. The details of the CLSγ-ray source and examples of nuclear experiments are described in the previous papers [7-9]. The storage ring NewSUBARU was operated with a mode of "Top-up", in which electron bunches were supplied continuously from an electron linac without extra acceleration into NewSUBARU. The energy of generated γ-rays is determined by the wave length of laser and the energy of electron beam. The maximum energy of the generated γ-ray beam was 16.7 MeV. The wave length of the laser was 1064 nm. The energy of the electrons was 974MeV. In this experiment, we used pulsed electron beams at a repetition rate of 2.5 MHz with a pulse width of 60 ps. A Q-switch Nd:YVO$_4$ laser system provided laser with a pulse width of 8 ns and a power of 4 W. The energy spread of the γ-ray beam depends on the collimator size and the emittance of the electron beam. We used two collimators as shown in Fig. 2. The first collimator was located inside of the storage ring. The second collimator was located in front of the target. The γ-ray energy width was about 5 MeV for Au and 3 MeV for NaI and Cu. The estimated γ-ray flux was (1-2)×10^6 photons/s in an energy range from 12 MeV to 16.7 MeV.

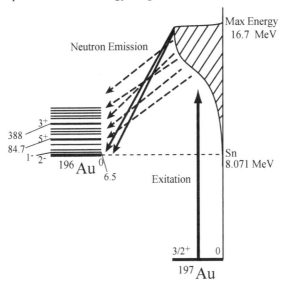

Fig 1. A partial level scheme of ^{196}Au, and a schematic view of the photonuclear reaction on ^{197}Au.

We irradiated the Au, NaI, and natural Cu targets. Their sizes were 10 mm$^\phi$ × 40 mm, 10 mm$^\phi$ × 50 mm, and 5 mm$^\phi$ × 50 mm for Au, NaI, and Cu, respectively. The second collimator with a diameter of 6 mm was located before the target position. The diameter of the incident beam was about 6 mm on the target NaI or Cu. This collimator was not used for the Au measurement and the beam diameter was larger than the Au target diameter.

The neutrons were measured using a time-of-flight (TOF) method. The detector was set at a polar angle of Φ = 90°outside the experimental room. A lead shield with a thickness of 2 mm was located in front of the detector. Each target was located inside the irradiation room with a concrete shield with a

304

thickness of 540 mm. Neutrons were guided to the detector through a hole with a diameter of 80 mm. A flight length was 970 mm.

A time-to-amplitude converter (TAC) was used to measure a time interval between time of the LCS γ-ray irradiation and time of neutron detection. A start signal is generated from an output of neutron detector. A stop signal was generated from the synthesizer signal with 500 MHz by the divider. This signal was also used as the external trigger signal to generate the laser pulse with 25 kHz. The synthesizer controlled the frequency of electron bunch with 2.5 MHz. The generation of the LCS γ-ray was synchronized with the electron bunch. The TAC signals were recorded using a multi-channel analyzer (MCA). The dead time of the MCA was below 3% and the neutron counting rate was 0.7-1.0 counts/s.

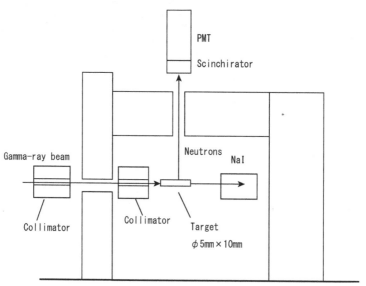

Fig. 2. Schematic view of experimental setup.

A key point of this experiment is how to determine the angle of linear polarization of the incident LCSγ-ray beam. The angle of the linear polarization plane was tuned by changing the linear polarization plane of the incident laser. The laser beam was extracted to the outside of the electron storage ring after the γ-ray generation and the polarization angle was measured. We measured the neutron energy spectra as a function of the laser polarization angle in a range from φ = 0° to 360° in 30° steps for NaI and Cu, where the φ = 0 was defined as the electric polarization vector being in the plane of the detector. We measured only 9 angles for the Au measurement to obtain high statistics.

3. Experimental Result

Figure 3 shows a TOF spectrum from the ^{197}Au measurement. There are a γ-ray peak and a neutron bump. The neutrons and prompt γ-rays are clearly separated as shown in Fig. 3. The measured time width of the prompt γ-rays is about 3 nswhich is longer than the expected duration of 60 ps from the electron pulse width because of time fluctuation of the slow rising-time photomultiplier and the time jitter between the laser generation time and the external trigger signal from the electron storage ring. The energy spectra of the neutrons were derived from the TOF signals. The measured maximum energy of the neutron is almost 8 MeV for Au. This energy is consistent with the energy of 8.6 MeV, the difference between the maximum photon energy of 16.7 MeV and the neutron separation energy of 8.1 MeV (see Fig. 2). The neutrons with energies lower than 2 MeV were not measured because of the detection efficiency. The time resolution, originated from the ratio of the detector depth to the flight pass length, is about 5%. Thus, the level structure of the residual nucleus cannot be observed in the present TOF spectra.

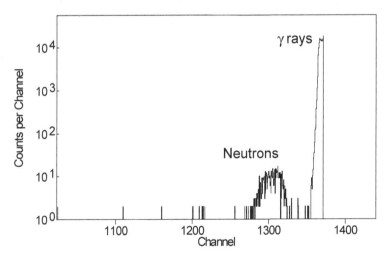

Fig. 3. A Time-Of-Flight spectrum for the Au target. The peaks corresponding to γ-rays and neutrons are clearly separately.

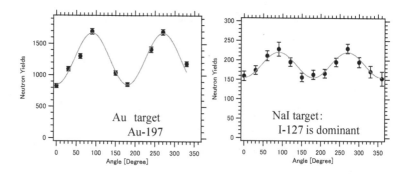

Fig. 4. Angular distribution for neutron cross sections in the $(\vec{\gamma}, n)$ reaction on Au and NaI targets. The solid lines are fitted using a function of $a + b\cos(2\phi)$.

4. Discussion

The neutron yields are presented as a function of azimuthal angle ϕ for An and NaI (see Fig. 4). The solid lines in Fig. 4 show functions in form of $a + b\cos(2\phi)$ obtained by χ^2-fitting. The NaI target includes ^{23}Na and ^{127}I. The neutron separation energies of these nuclides are 12.4 MeV (^{23}Na) and 9.1 MeV (^{127}I) and the contribution of ^{127}I is dominant for the NaI target. The experimental result for Cu is also reproduced by the function $a + b\cos(2\phi)$ although the anisotropy, b/a, is very low. The three neutron angular distributions are described as the function of $a + b\cos(2\phi)$ independent of nuclides. Therefore, the theoretical prediction by Agodi is for the first time verified in the wide mass region.

The targets used in the present experiments are odd-A isotopes. The residual isotope is an odd-odd nucleus. The level density of the odd-odd nuclei is higher than even-even nuclei and odd-A nuclei. There are many combinations of the transitions as shown in Fig. 1. Therefore, it is difficult to calculate realistic values for neutron angular distributions for the odd-A isotopes.

In contrast, in the case of an even-even nucleus target, the spin and parity of states populated by the incident photon is in general limited only to $J^{\pi}= 1^{\pm}$ and 2^{+} since E1, M1, and E2 transitions are dominant. Thus, the theoretical calculation is easy comparing with those of odd-A targets. We have derived an explicit form of E1, M1, and E2 transitionmatrix elements from a $J^{\pi}=0^{+}$ target nucleus, with the long wave-length approximation to the photon field. As shown by Agodi, both E1 and M1 transitions give the ϕ dependence of $a + b\cos(2\phi)$ of the neutron emission cross section induced by a linearly polarized photon beam.

5. Conclusion

Agodi predicted theoretically the anisotropic angular distribution of neutrons emitted from states excited via dipole transitions with linearly polarized γ-ray beam at the polar angle of $\Phi=90°$ should be followed by a simple function, a + bcos(2ϕ), but it has not been verified experimentally over half a century. We have measured neutron angular distribution using linear polarizedγ-ray beam generated by Compton scattering at NewSUBARU. We have verified this Agodi's prediction for the first time in the wide mass region.

References

1. W. Bertozzi, R. Ledoux, Nucl. Inst. Meth. Phys. Res. B 241, 820 (2005).
2. J.Pruet *et al.*, Appl. Phys. 99, 123102 (2006).
3. R. Hajima *et al.*, J. Nucl. Sci. Technol. 45, 441(2008).
4. N. Kikuzawa *et al.*, Appl. Phys. Exp. 2, 036502-1 (2009).
5. T. Hayakawa *et al.*, Nucl. Inst. Method. Phys. Res. A 621, 695 (2010).
6. A. Agodi, Il Nuovo Cimento 5, 21(1957).
7. S. Miyamoto *et al.*, Radiation Measurements 41, S179 (2007).
8. T. Hayakawa *et al.*, Phys. Rev. C 74, 065802 (2006).
9. T. Hayakawa *et al.*, Phys. Rev. C 77, 068801 (2008).

Test Experiment of γ-Ray Diffraction for Crystal Monochromators

S. Matsuba*, C. T. Angell, T. Shizuma, T. Hayakawa, N. Nishimori, R. Nagai, M. Sawamura, and R. Hajima

Quantum Beam Science Directorate
Japan Atomic Energy Agency,
Tokai, Ibaraki 319-1112, Japan
matsuba.shunya@jaea.go.jp
www.jaea.go.jp

A γ-ray diffraction property of a silicon single crystal was studied in the Laue geometry using 1.33 MeV and 1.17 MeV γ-ray of ^{60}Co. The thickness was chosen to maximize the reflectivity of (440) lattice plane in γ-ray energy ranges of 1 to 2 MeV. We measured diffracted γ-rays from the crystal by an NaI scintillator. A measured diffraction intensity of 1.33 MeV γ-rays was 18 counts/sec by using 3.0 TBq a ^{60}Coγ-ray radiation system. The measured intensity is 30 % lower than an expected value.

Keywords: Monochromator; γ-ray.

1. Introduction

Next generation γ-ray sources on basis of laser Compton scattering (LCS) are developed for nondestructive isotope identification and other application for basic science [1, 2]. LCS γ-rays are produced by collision of laser photons and high energy electrons in an accelerator. The brightness of γ-rays which mainly depends on electron beam emittance will be several orders of magnitude higher than existing LCS γ-ray sources based on storage rings when low emittance electron beams are produced by energy recovery linac. Bragg diffraction by perfect crystals is suitable for highly monochromatization in such high brightness γ-rays.

Crystal monochromators were not used generally in MeV region γ-ray except Institute Laue Langevin (ILL) [3], although a resolution of crystal monochromators is higher than that of high-purity germanium detector. This is because efficiency of crystal monochromators is too low and resolution of high-purity germanium detectors are sufficient enough for many purposes. However, according to appearance of high brightness γ-ray sources, the attention for crystal monochromators has been growing.

The crystal monochromators in ILL are used in combination with neutron capture γ-rays. Here we propose to use crystal monochromators together with LCS γ-ray. As the first step we studied the γ-ray diffraction by combining

310

the ^{60}Co source and a Si crystal. In this paper we report on our recent results.

2. Experimental Setup

Figure 1 shows an experimental setup using 1.17 MeV and 1.33 MeV γ-rays from a ^{60}Co source in the Facility of Radiation Standard of Japan Atomic Energy Agency. The ^{60}Co intensity is about 3.0 TBq. Horizontal angular spread of γ-ray is approximately 2.5 mrad determined by the second collimator. Vertical angular spread is about 6.7 mrad determined by the third collimator. An NaI scintillator with a size of 2 inch diameter and 2 inch length was used to detect diffracted γ-rays.

Bragg diffraction occurs when Bragg's law is satisfied. Bragg's law is given by

$$2d \sin \theta_B = n \cdot \lambda$$

where d is the lattice plane spacing of the crystal, θ_B is the Bragg angle, n is the order of diffraction, λ is the wavelength of γ-ray. The 2mm thickness of the Si crystal was optimized for 1.5 MeV γ-ray diffraction from (440) lattice plane in the Laue geometry. The Bragg angles of Si (440) are 4.84 mrad for 1.33 MeV γ-ray and 5.50 mrad for 1.17 MeV. The acceptance angle of these conditions is estimated to be about 50 nrad from the dynamical theory of diffraction [4].

Diffracted γ-rays intensity is determined by a γ-ray flux on the crystal and the ratio between acceptance angles and a beam divergence. The intensity is expected to be approximately 30 counts per seconds.

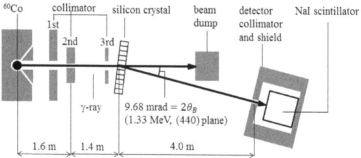

Figure 1. The Experimental setup of MeVγ-ray diffraction from a silicon crystal. Three lead collimators areplaced in front of the silicon crystal. The first and third collimators have ahole with 10 mm wide and 20 mm height opening. Their thicknesses are 20 cm and 5 cm, respectively. The second collimator with a thickness of 20 mm has a hole with 4 mm wide and 20 mm height. The detector lead collimator with a thickness of 15 mm has a hole with 4mm wide and 40 mm height.

2.1. Flux optimization

Before diffraction measurements, the γ-ray flux at the position of the crystal was measured and optimized by moving the second collimator. A setup of flux measurements is shown in Fig. 2. A CeBr3 scintillator with a size of one inch diameter by one inch length was replaced instead of the silicon crystal. The threshold of the discriminator was set at the γ-ray energy of 1.17 MeV. In order to avoid the counting saturation, an 8.5 cm thick lead absorber was inserted in front of the detector to reduce the γ-ray intensity.

First, the γ-ray flux was measured without collimators. The results were 4.18×10^5 photons/sec. This value is the same as the estimated value from mass attenuation, buildup factor, and solid angle within error.

Next, two collimators were set. The γ-ray flux was maximized by shifting and rotating the second collimator. Figure 3 shows the results. The collimator was moved about 0.5 mm and 0.1 degree from the initial position.

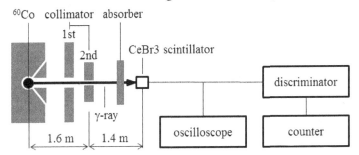

Figure 2. The setup of flux measurements.

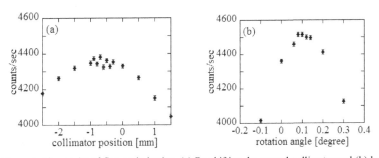

Figure 3. The results of flux optimization. (a) By shifting the second collimator and (b) by rotating the second collimator. Note that we still do not understand the reason why the data have a zig-zag pattern around the peak position in the left figure. The issue does not affect conclusions.

3. Diffraction Measurements

After the flux measurements, diffraction measurements were carried out. The crystal was arranged to make the polished surface orthogonal to the beam axis. Since, the polished surface was deviated from a lattice plane, we expected that there happened some misalignment between the beam axis and the lattice plane.

First, a crystal angle which guided the diffraction beam to the detector was searched for. Since the beam divergence was wide about 2.5 mrad, the diffraction was expected to occur at the same range of the crystal rotation. The acceptance of the detector collimator was about 1 mrad that was determined by the distance (4 m) and the hole width(4 mm). Therefore,γ-ray intensities were measured with a scan step of 0.5 mrad until the diffractedγ-rays were observed. The results indicated that the difference angle between polished surface and lattice plane was 2.1 mrad.

After the above mentioned procedures, a crystal angle dependence of γ-ray intensities was measured in more detail. Since the γ-ray detection intensity was determined by convolution of diffraction intensity and acceptance of the detector collimator. The detected intensities of 1.33 MeV and 1.17 MeV γ-rays dependon the crystal angle.

At the measurements, the crystal was rotated in the range of± 5 mrad. When an angle between the crystal lattice plane and the beam axis was close to 4.84 mrad (Bragg angle),γ-rays were diffracted by the Si crystal and entered the detector. Figure 4 shows spectra measured by the NaI scintillator. In the case that the Si crystal lattice plane angle is different from the Bragg angle,γ-rays scattered before crystal collimators were detected. The γ-ray counts increased when the glancing angle of the lattice plane approached to 4.84 mrad,

First, the intensity of 1.33 MeV γ-rays is obtained by summing the counts of the full energy peak. Next, 1.17 MeV γ-ray intensity was analyzed. The full energy peak of 1.17 MeV was observed on top of the slope due to the Compton background from 1.33 MeV γ-rays. Therefore the background is subtracted by assuminga straight line.

The crystal angle dependence of the γ-ray intensities were shown in Fig. 5. The crystal angle dependence was varied as a function of γ-ray energies, that is derived from the difference of Bragg angle. The maximum intensity of 1.33 MeV γ-rays was about 6.3 counts per seconds. The detection efficiency of the NaI scintillator was about 30 % for 1.33 MeV γ-rays. Therefore, the diffracted γ-ray intensities were estimated to be 21 counts per seconds that is 30 % lower than that of the expected value.

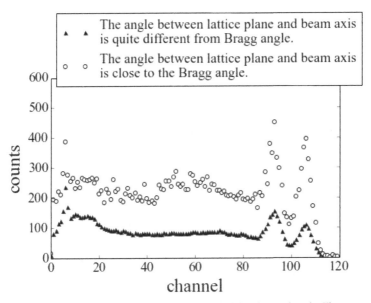

Figure 4. γ-ray spectra measured with two different setups of the Si crystal angle. The measurement time of both the spectra is 300 seconds.

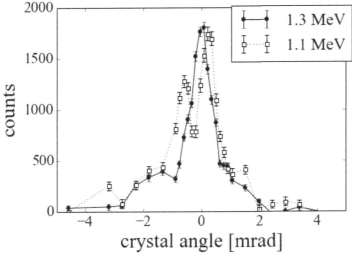

Figure 5. The crystal angle dependences of the observed γ-ray intensity. The zero degrees of x-axis is set at the angle where the counts of 1.33 MeV γ-rays maximize.

4. Summary

1.33 MeV and 1.17 MeV γ-rays from ^{60}Co were diffracted by the crystal, and were measured by the NaI scintillator for the development of crystal monochromators used together with LCS γ-ray in MeV region.

The crystal angle dependences of diffracted γ-ray intensity were obtained. The dependences were changed as a function of γ-ray energy because of Bragg angle variation. The diffractedγ-ray intensity was estimated to be 21 counts per seconds that is 30 % lower than that of the expected value.

In future, we will study the detailed crystal angle dependence for the γ-ray diffraction.

Acknowledgments

We are grateful to S. Nishino and other members of Facility of Radiation Standards for their support of our experiments. We would like to thank M. Fujiwara for his advice.

References

1. R. Hajima, N. Kikuzawa, N. Nishimori, T. Hayakwa, T. Shizuma, K. Kawase, M. Kando, E. Minehara, H. Toyokawa and H. Ohgaki, Detection of radioactive isotopes by using laser Compton scattered g-ray beams,*Nucl. Instrum. and Meth.* A608 S57 (2009) .
2. The ELI-Nuclear Phsics working groups, The White Book of ELI Nuclear Physics, (Bucharest-Magurele, Romania, 2009).
3. E. G. Kessler Jr., M. S. Dewey, R. D. Deslattes, A. Heins, H. G. Borner, M. Jentschel, H. Lehmann, *Nucl. Instrum. and Meth.* A457 187 (2001).
4. W. H. Zachariasen. Theory of X-ray diffraction in crystals. (John Wiley and Sons, New York, 1945).

Overview of Laser Compton-Scattered Photon Source at the cERL

R. Nagai*, R. Hajima, A. Kosuge, M. Mori, T. Shizuma and N. Nishimori

Quantum Beam Science Directorate,
Japan Atomic Energy Agency,
Tokai, Ibaraki 319-1112, Japan
*nagai.ryoji@jaea.go.jp
www.jaea.go.jp

T. Akagi, Y. Honda and J. Urakawa

1High Energy Accelerator Research Organization (KEK),
Tsukuba, Ibaraki 305-0801, Japan
www.kek.jp

A high intensity γ-ray source from the laser Compton scattering (LCS) by an electron beam in the energy recovery linac (ERL) is a very useful prove for a nondestructive assay to identify nuclear species. In order to demonstrate a high performance of the accelerator and laser required for the γ-ray source, an LCS experiment is planned at the Compact ERL (cERL) at High Energy Accelerator Research Organization (KEK). A mode-locked fiber laser, laser enhancement cavity, beamline, and experimental hatch are under construction for the LCS experiment.

Keywords: Laser compton scattering; energy-recovery linac; compact ERL.

1. Introduction

A nondestructive assay system to identify nuclear species by means of nuclear resonance fluorescence (NRF) is under development in Japan Atomic Energy Agency (JAEA) for the nuclear nonproliferation and security. Quasi-monochromatic γ-rays are generated by laser Compton scattering (LCS) based on energy-recovery linac (ERL) accelerator and laser technologies. In case of the LCS γ-ray source, the γ-ray energies can be selected by changing the electron energy, laser wavelength or collision angle. Furthermore, the energy width of γ-rays can be sharpened by putting a small diameter collimator to restrict the scattering angle. The LCS γ-ray source has specific different characteristics from other conventional γ-ray sources in its energy tunability, narrow energy width and small divergence.

For the generation of high-flux and high-brightness γ-rays by LCS, a small-emittance and high-current electron beam and a high-power laser are necessary. An ERL is the optimum apparatus to accelerate electron beams with small emittance and high-average current[1]. The Compact ERL (cERL)[2] at High Energy Accelerator Research Organization (KEK) has been constructed as a test accelerator for ERL-based light sources. In order to demonstrate the performance of the accelerator combining with laser for generating γ-rays, an LCS experiment is planned at the cERL. A mode-locked fiber laser, laser enhancement cavity, beamline, and experimental hatch are under construction for the LCS experiment.

2. Overview of the Compact ERL

The performance of ERL, e.g., electron beam current and emittance, is restricted by its electron source and accelerating structure. The photocathode electron gun and superconducting accelerator (SCA) are key technologies to accelerate a small emittance and high-current electron beam. The cERL has been constructed to demonstrate the manifestation of these key technologies. Design parameters of the cERL are listed in Table 1. The layout of the cERL is shown in Fig. 1. Since the design energy of electron beams at the cERL is 35 MeV, LCS photons will have an energy of 22 keV. However, the LCS experiment at the cERL is regarded as a demonstration of high-flux LCS γ-ray source for the nondestructive assay system to identify nuclear species. Increasing the LCS photon energy is simply achievable by adding more SCAs for higher electron energy. A space for additional SCAs and second recirculation loop is reserved for a future energy upgrade of the cERL[3] as shown in Fig. 1. After the full upgrade of the cERL, an electron beam of 245 MeV will be available and will produce 1 MeV γ-rays.

A DC electron gun equipped with a semiconductor photocathode has been designed and fabricated at JAEA to generate small-emittance and high-current electron beams. The beam generation of the electron gun has been successfully demonstrated at the world-highest voltage, 500 kV, and a high-average current up to 10 mA[4]. In case of usual high-voltage DC photocathode guns, the operational voltage has been restricted to 350 kV or lower owing to the field emission problem, which causes electrical breakdown or punch-through on the ceramic insulator. In case of the JAEA gun, a segmented insulator with guard-rings has been employed to keep the insulator safe from the field emission generated from a central stem electrode to solve the field emission problem[5]. After the successful demonstration of 500-keV beam generation, the gun was shipped to KEK, and was installed

Table 1.   Design parameters of cERL.

Beam energy [MeV]	35 (initial goal)
	245 (upgradable in future)
Injection energy [MeV]	5
Beam current [mA]	10 (initial goal)
	100 (future goal)
Normalized emittance [mm mrad, rms]	1 (initial goal)
	0.1 (at low current)
RF Frequency [MHz]	1300
Bunch length [ps, rms]	1–3 (usual)
	≤ 0.1 (under compression)

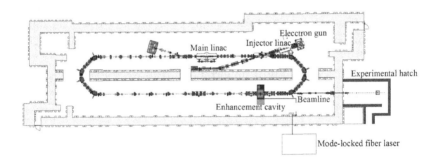

Fig. 1.   Layout of the cERL.

at the cERL. The beam operation of the cERL was started from April 2013. At present, the gun provides an electron beam for daily operation of cERL.

Two types of SCAs for the cERL have been developed at KEK, one for the injector and the other for the main linac[6]. The operation frequency and temperature of the SCAs are 1300 MHz and 2 K, respectively. The injector is three 2-cell cavities housed in the cryomodule. In the injector, a high-current electron beam is accelerated without energy recovery. Therefore, a twin-coupler system was employed to reduce the input power per coupler. The main linac is two 9-cell cavities housed in the cryomodule. In the main linac, suppression of higher-order mode (HOM) is a main issue to avoid the beam-breakup instability. A beampipe type HOM absorber was employed to achieve strong HOM damping. The HOM absorbers are placed at both the ends of the 9-cell cavity and inside the cryomodule, at temperature of

80 K. The electron beam generated in the photocathode gun is accelerated to 5 MeV by the injector. The beam is further accelerated to 35 MeV by the main linac and radiate the γ-rays at the LCS section. The beam injected again to the main linac through the recirculation loop is deccelated to 5 MeV, and is dumped. The commissioning of the cERL was started from December 2013. The energy recovery operation was successfully achieved in February 2014.

3. Equipment for the LCS experiment

In order to demonstrate high-flux and high-brightness LCS beam generation at the cERL, a laser enhancement cavity will be installed at the recirculation loop of the cERL. The LCS beam is transported to an experimental hatch through a vacuum evacuated beamline for evaluation of the LCS beam. Figure 1 shows a floor layout of the equipment of the LCS experiment at the cERL. The installation of the equipment at the cERL will be started from the summer in 2014. The LCS experiment is scheduled in March 2015.

Since the cross-section of the Compton scattering is small, reusing the laser photons efficiently is important to realize a high-flux and high-brightness γ-ray source. Efficient reuse of laser photons is achieved by introducing a laser enhancement cavity. The laser enhancement cavity is a high-finesse Fabry-Perot optical cavity, which stores optical pulses injected from an external mode-locked laser. In the experiment, a 4-mirror cavity is employed to achieve high stability and small waist size[7]. As shown in Fig. 2, two cavities are stacked in the same gimbals. Design parameters of the enhancement cavity are listed in Table 2.

Table 2. Design parameters of the enhancement cavity.

Frequency [MHz]	162.5
Enhancement factor	2550
Collision angle [deg]	18
Spot size [μm, rms]	20 (hor.), 30 (vert.)

A high-power mode-locked fiber laser for the LCS experiment is under development at Kansai Photon Science Institute, JAEA[8]. The laser consists of a mode-locked oscillator and 4-stage amplifiers; all of them utilize Yb-doped fibers as laser gain media. Since the LCS γ-ray bandwidth is affected by laser bandwidth, a high-power laser with narrow-bandwidth was

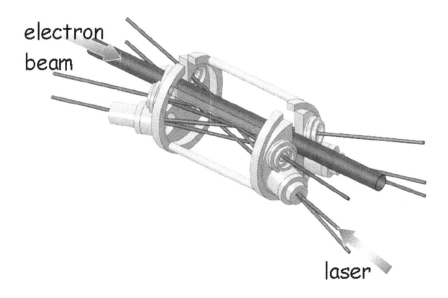

electron
beam

laser

Fig. 2. Schematic drawing of the enhancement cavity.

developed. The laser system is, thus, equipped with two pulse stretchers and one compressor to avoid nonlinear spectral broadening during the amplification. Bandpass filters are also inserted between amplification stages. Design parameters of the mode-locked fiber laser are listed in Table 3

Table 3. Design parameters of the mode-locked fiber laser.

Average power [W]	~ 100
Center wavelength [nm]	1030
Band width [nm, FWHM]	1.4
Pulse duration [ps]	~ 2
Repetition rate [MHz]	162.5

4. Summary

A quasi-monochromatic γ-ray source based on an ERL accelerator and LCS is under developed for a nondestructive assay system. In order to demonstrate the accelerator and laser performance required for the γ-ray source, an LCS experiment is planned at the cERL, which is a test accelerator for ERL-based light sources. The energy recovery operation at the cERL has been achieved in February 2014. A mode-locked fiber laser, laser enhancement cavity, beamline, and experimental hatch will be installed at the cERL for the LCS experiment. The LCS experiment is scheduled in March 2015.

References

1. R. Hajima, *et al.*, Rev. Acc. Sci. and Tech. **3**, 121–146 (2010).
2. S. Sakanaka, *et al.*, Proc. of IPAC2013, 2159–2161 (2013).
3. M. Shimada, *et al.*, Proc. of IPAC2011, 1909–1911 (2011).
4. N. Nishimori *et al.*, Appl. Phys. Lett. **102**, 234103 (2013).
5. R. Nagai *et al.*, Rev. Sci. Instr. **81**, 033304 (2010).
6. K. Umemori, *et al.*, Proc. of SRF2009, 896–901 (2009).
7. T. Akagi, *et al.*, Proc. of IPAC2012, 2645–2647 (2012).
8. M. Mori, *et al.*, Proc. CLEO-PR-2013 and OECC-2013, paper:MD1-4 (2013).

Development of a High-Brightness and High-Current Electron Gun for High-Flux γ-Ray Generation

N. Nishimori*, R. Nagai, S. Matsuba and R. Hajima

Japan Atomic Energy Agency (JAEA)
Tokai, Ibrakaki 319-1195, Japan
*nishimori.nobuyuki@jaea.go.jp

M. Yamamoto, Y. Honda, T. Miyajima and T. Uchiyama

High Energy Accelerator Research Organization (KEK)
Oho, Tsukuba, Ibaraki 305-0801, Japan

M. Kuriki

Hiroshima University
Higashihiroshima, Hiroshima 464-8603, Japan

A high-flux mono-energetic γ-ray beam can be generated via Compton scattering of high-power laser by high-brightness electron beam. We have developed a high-brightness and high-current electron gun for generation of the high-flux γ-ray beam. Recently we demonstrated 500 keV electron beam generation, which meets the high-brightness requirement, from our DC photocathode gun at Japan Atomic Energy Agency. The gun was transported to High Energy Accelerator Research Organization (KEK) and connected to the following accelerator system. The gun operational status at KEK and our plan to develop a multialkali photocathode with a long lifetime are presented.

Keywords: Photocathode; electron gun; energy recovery linac.

1. Introduction

The quantification of fissile materials such as uranium-235 and plutonium-239 in spent fuel assemblies requires isotope-specific identification. We propose the use of nuclear resonance fluorescence (NRF) to identify the isotopic composition of sample materials in a nondestructive way[1,2]. A mono-energetic γ-ray beam tuned to the resonance state of the target nuclear isotope is injected into the spent fuels, and NRF γ-rays generated spherically from the fuels are detected with a γ-ray detector. The proposed nondestructive isotope identification system requires a high-intensity mono-energetic γ-ray beam. Although a mono-energetic γ-ray beam can be generated using a conventional laser Compton scattering technique, the generation of a high-intensity γ-ray beam requires an electron beam of unprecedentedly high brightness, which can be generated with an advanced accelerator system known as the energy recovery linac (ERL). One of technological challenges of the ERL system is the development of a high-brightness, high-current electron gun.

We have developed a 500-kV DC gun[3] for the future ERL systems at Japan Atomic Energy Agency (JAEA). Recently we demonstrated generation of a 500-keV electron beam from the DC photocathode gun[4]. This demonstration was achieved by solving two electrical discharge problems. One is discharge on the insulator's ceramic surface caused by field emission generated from a central stem electrode. We solved this problem by employing a segmented insulator with rings to protect the insulator from the field emission[5]. The other is microdischarge at an anode electrode or a vacuum chamber, which is triggered by microparticle transfer or field emission from a cathode electrode. Our experimental investigation revealed that a larger acceleration gap, optimized mainly to reduce the surface electric field of the anode electrode, suppresses the microdischarge events that accompany gas desorption. It was also found that non-evaporable getter pumps placed around the acceleration gap greatly help to suppress those microdischarge events[6].

The electron gun developed at JAEA was transported to the compact ERL (cERL) at High Energy Accelerator Research Organization (KEK) in October 2012. The cERL has been developed to demonstrate reliable operation of key components of future ERL machines[7]. In this paper, we present operational status of our electron gun at the cERL and our plan to develop a multialkali photocathode as an alternative for GaAs photocathode, which is currently used for beam generation.

2. Gun operation at the cERL

The electron gun installed at the cERL is followed by an injector super-conducting accelerator (SCA), a main SCA, an energy recovery loop, and a beam dump. A laser super-cavity will be installed in the energy recovery loop to generate a high-flux γ-ray beam for proof-of-principle (PoP) experiment of nondestructive isotope identification. The current status of the PoP experiment is described elsewhere[8].

After installation to the cERL, it turned out that two out of ten segments of the segmented insulator had problems on their surfaces. The

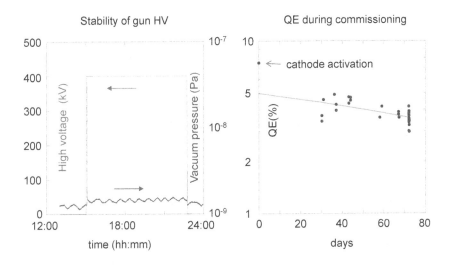

Fig. 1. The left shows the gun high voltage stability and the gun vacuum pressure during beam commissioning over eight hours at the cERL. The right shows QE of GaAs photocathode during beam commissioning. The 1/e lifetime is obtained as 5000 hours.

insulator ceramic is thus operated with eight segments, which limits the maximum applied voltage around 440 kV instead of 550 kV. The gun was high voltage conditioned up to 420 kV and has stably been operated at 390 kV without any breakdown for 200 hours during the injector commissioning. The left of Fig. 1 shows the gun high voltage stability and the vacuum pressure during the gun operation for eight hours. The gun vacuum pressure was 1.4×10^{-9} Pa when connected to the downstream beam line. A GaAs wafer activated with cesium and oxygen has been used as a photocathode in our gun system. The quantum efficiency (QE) of GaAs photocathode has been obtained from the measurements of beam current and laser power, as shown in Fig. 1. The typical QE was 4% and its 1/e lifetime was measured to be 5000 hours. The GaAs photocathode was used during the commissioning without any reactivation. This indicates that the photocathode performance is good enough for commissioning purpose with a low current beam.

3. Alkali photocathode development

We have developed a multialkali photocathode inspired by demonstration of high current electron beam generation at Cornell University[9]. The

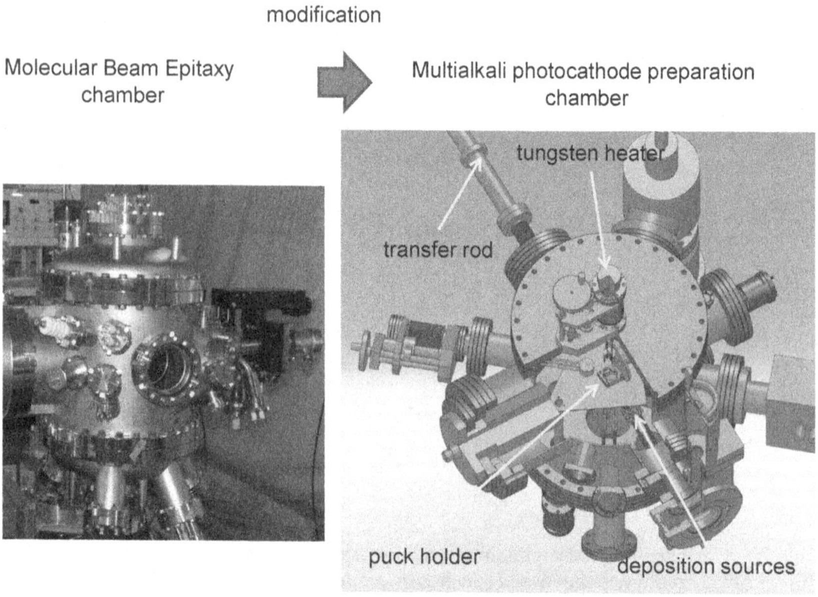

324

modification

Molecular Beam Epitaxy
chamber

Multialkali photocathode preparation
chamber

Fig. 2.   Existing molecular beam epitaxy chamber (left) will be modified to a multialkali photocathode preparation chamber (right).

multialkalai photocathode performance is superior in temporal response and lifetime to GaAs. It is inferior in emittance, but the emittance difference is only 10 - 20 % at the laser wavelength of 530nm, which is usually used for unpolarized beam generation. Thus the multialkali photocathode recently draws attention in photocathode gun community.

The existing molecular beam epitaxy chamber will be modified to accommodate the multialkali photocathode preparation. We designed a preparation system consisting of five deposition sources to fabricate photocathodes with various combination of alkali materials. The photocathode fabricated in the preparation chamber can be transferred to our electron gun system for beam generation. Figure 3 shows a 250 kV - 50 mA photocathode DC gun system. This system was originally developed for beam generation with a GaAs photocathode. We will connect the new multialkali photocathode preparation system to the gun, and perform beam generation test with an existing beam line.

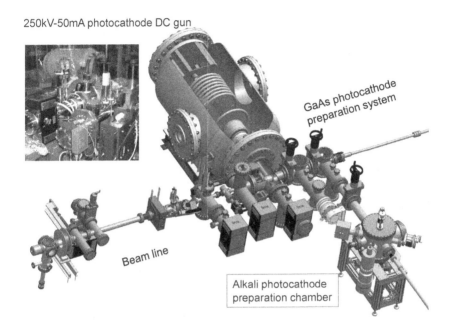

250kV-50mA photocathode DC gun

GaAs photocathode preparation system

Beam line

Alkali photocathode preparation chamber

Fig. 3. A 250 kV - 50 mA photocathode DC gun connected to the alkali photocathode preparation chamber which is under development. The existing 250 kV gun will serve as a test stand to measure the performance of the multialkali photocathode.

4. Summary

A high-brightness, high-current electron gun has been developed. We successfully generated 500 keV electron beam from the gun, satisfying the high-brightness requirement. The gun has stably provided an electron beam for the cERL commissioning. The high-current performance requires a long lifetime photocathode. To meet this requirement we started to develop a multialkali photocathode.

Acknowledgments

This work was supported by Photon and Quantum Basic Research Coordinated Development Program from the Ministry of Education, Culture, Sports, Science Technology, Japan, and partially supported by a JSPS Grant-in-Aid for Scientific Research in Japan (23540353).

References

1. R. Hajima et al., J. Nucl. Sci. Technol. **45**, 441 (2008).
2. T. Hayakawa et al., Nucl. Instr. and Meth. A **621**, 695 (2010).
3. N. Nishimori et al., J. Phys.: Conf. Ser. **298**, 012005 (2011).
4. N. Nishimori et al., Appl. Phys. Lett. **102**, 234103 (2013).
5. R. Nagai et al., Rev. Sci. Instrum. **81**, 033304 (2010).
6. N. Nishimori et al., submitted to Phys. Rev. STAB.
7. S. Sakanaka et al., "Construction and Commissioning of Compact-ERL Injector at KEK", in Proc. of ERL2013, 17 (2013).
8. R. Nagai et al.,"Overview of the equipment for a demonstration of laser Compton-scattered photon source at the cERL", in these proceedings.
9. B. Dunham et al., Appl. Phys. Lett. **102**, 034105 (2013).

Design of ERL Spoke Cavity
For Non-Destructive Assay Research

M. Sawamura*, R. Nagai, N. Nishimori and R. Hajima

Gamma-ray Non-Destructive Assay Research Group
Japan Atomic Energy Agency
Tokai, Ibaraki 319-1195, Japan
**Sawamura.masaru@jaea.go.jp*

We are proposing non-destructive assay system of nuclear materials with laser Compton scattering combined with an energy-recovery linac (ERL) and a laser. Since constructing accelerator system for nuclear safe guard and security requires small cavities, spoke cavities have many advantages such as shortening the distance between cavities, small frequency detune due to micro-phonics and easy adjustment of field distribution for strong cell coupling.

Calculations of optimized cavity shape and HOM coupler shape have been performed and rf properties with aluminum spoke cavity model have been also measured. Considering refrigerator system required for superconducting accelerator, we are planning to develop 325MHz spoke cavity which can be practically operated with 4K liquid helium. We have started to fabricate the niobium one-spoke cavity.

Keywords: Superconducting accelerator; spoke cavity; energy-recovery linac.

1. Introduction

Low emittance and high current beam from Energy Recovery-Linac (ERL) can generate high brightness and high quality light source. The high quality beam of ERL can also significantly improve brightness and monochromaticity of X/γ-ray generated by laser Compton scattering (LCS). Nuclear resonance fluorescence (NRF) with the LCS-γ ray can be utilized to nondestructively inspect nuclear materials such as uranium, plutonium and minor actinoid elements in spent reactor fuels. This method is significant technology for nuclear safeguards and security. We are proposing non-destructive assay system of nuclear materials by LCS combined with an ERL and a laser [1].

Practical use of this system requires downsizing the ERL so that it is important to compact the accelerating cavity. Since the beam instability due to higher-order modes (HOMs) limits the beam current of the ERL, HOM damping is significant for the ERL cavities. HOM damping devices such as HOM absorbers and HOM couplers for elliptical cavities tend to increase the total accelerator length since they are attached to the beam pipes. On the contrary, spoke cavities have an advantage of shortening the total accelerator length.

Optimization of the spoke cavity shape, HOM coupler design, and mechanical design of the spoke cavity have been performed. RF properties for the spoke cavity model were also measured. The present paper describes these results.

2. Advantages of Spoke Cavity

The superconducting spoke cavity used for ERL has following advantages.

1) A superconducting cavity requires HOM absorbersor HOM couplers to damp HOM and input couplers to feed RF power into the cavity. These elements can be installed along the side of the spoke cavity so that the total length of spoke cavity can get shorter than that of elliptical cavity and the distance between the cavities can be decreased.

2) The resonant frequency of spoke cavity mainly depends on the spoke length, and high cavity stiffness reduces the fluctuation of cavity resonant frequency due to microphonics. The small frequency fluctuation of the ERL cavity can decrease the required RF power and tolerance of the input coupler. This results in making the RF power supply compact.

3) When the outer size of spoke cavity is similar to that of elliptical cavity, the resonant frequency of spoke cavity is nearly half of elliptical cavity. Lower frequency can decrease the energy spread because of the narrow accelerating phase spread for the same bunch length beam. Small energy spread beam can increase the brightness of LSC X/γ-ray.

4) Cell coupling of spoke cavity is stronger than that of elliptical cavity. Stronger cell coupling makes the field flatness easier to adjust and less disturbed to increase number of cells. This increases the effective accelerating length.

3. Status of Development of Spoke Cavity

3.1. *Design of Spoke Cavity Shape*

Since the spoke cavity shape is more complicated than the elliptical cavity shape and there are a few objectives to be optimized, multi-objective optimization using genetic algorithm is adopted to determine the spoke cavity shape. We deal with two objectives ofthe ration of maximum electric field to accelerating field (Epeak/Eacc), and the ration of maximum magnetic field to accelerating field (Hpeak/Eacc).

The Pareto fronts of Epeak/Eacc and Hpeak/Eacc in every 5 generations are shown in Fig. 1. With each succeeding generation Pareto fronts are improving and gradually approaching to the final Pareto front [2].

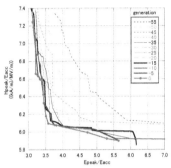

Fig. 1. The Pareto fronts of Epeak/Eacc and Hpeak/Eaccin every 5 generations.

3.2. Measurement of RF Properties of Spoke Cavity Model

An aluminum spoke cavity model was fabricated to measure RF properties. The main parameters of the model are shown in Table 1. Design of easy spoke replacement caused bad contact between spoke and tank to result in adverse effect of 60-80% Q values of calculation.

The resonant modes from 500MHz to 1500MHz were measured with a network analyzer as shown in Fig. 2. The monopole and dipole modes were determined by measuring the electric field distribution with bead pull method. There were no monopole modes around the double frequency of the accelerating mode or no dipole modes of higher Q value than the others which cause beam instability in ERL operation.

Table 1. Parameters of aluminum spoke cavity model.

Material	Aluminum
Frequency	650MHz
No. of spokes	2
No. of cells	3
Cell length	230.6 mm
Radius of cylinder	274mm
Bore radius	40mm

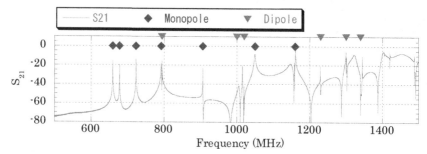

Fig. 2. Measured monopole and dipole modes spectrum of spoke cavity model.

3.3. Design of HOM Couplers

HOM damping is important when the spoke cavity is used for superconducting accelerator. The beam-line HOM damper cannot be used because of narrow beam pipe of spoke cavity. An HOM coupler can be attached on the side of the cavity. The similar HOM coupler used for an elliptical cavity was designed with the accelerating frequency of 650MHz as shown in Fig. 3. This type of HOM couplers become rather big and have problem of the heating at the connectors [3,4].

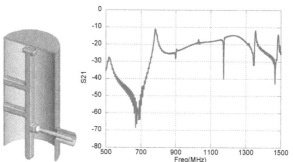

Fig. 3. Calculated transmit coefficients of conventional HOM couplers.

We propose a new type of HOM coupler which combines coaxial line and waveguide as shown in Fig. 4. Though all modes including the accelerating mode can transmit through the coaxial line, only HOM modes can transmit through the waveguide whose cutoff frequency is just above the accelerating frequency by choosing proper waveguide width. The calculated result of transmit coefficient is shown in Fig. 4. The transmit coefficient rapidly increases just above the accelerating frequency.

We also propose another new HOM coupler of C-shape waveguide by deforming this coaxial-waveguide HOM coupler as shown in Fig. 5. The cutoff frequency is determined by the average arc length of C-shape part as shown in Fig. 5. The input and output parts can be converted to coaxial lines. Since the inner conductor are connected to the outer conductor, the inner conductor can be efficiently cooled down. This can settle the heating problem of inner connector of HOM coupler.

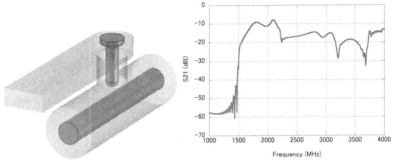

Fig. 4. Outline view (left) and calculated transmit coefficients (right) of coaxial-waveguide HOM coupler.

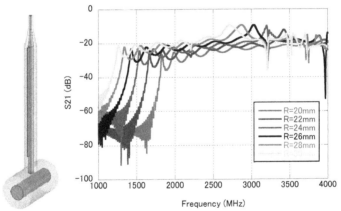

Fig. 5. Outline view (left) and calculated transmit coefficients (right) of coaxial- C-shape waveguide HOM coupler for various radii which change cutoff frequency depending on the average arc length.

3.4. *Mechanical Design of Spoke Cavity Model*

The strength calculation required for the fabrication of niobium cavity was performed. Messes stress under vacuum condition was calculated with simulation code of ABAQUS by changing the thickness of the wall. The

332

thickness of 3mm is enough except for the end plates and base of the spokes. The longitudinal support plates at the spoke base and the radial support plates at the end plate reduced the stress as shown in Fig.6.

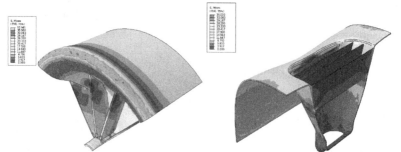

Fig. 6. Calculated stress distribution of spoke cavity under vacuum pressure at the end plate (left) and at the spoke (right).

4. Summary

We have been designing and measuring the cavity components for the spoke cavity. We are planning to fabricate the niobium spoke cavity and will check the performance of the cavity by vertical test.

Acknowledgments

This work was supported by Photon and Quantum Basic Research Coordinated Development Program from the Ministry of Education, Culture, Sports, Science and Technology, Japan.

References

1. R. Hajima et al., Proc. of 8th Annual Meeting of Particle accelerator Society of Japan, TUPS040, 2011
2. M. Sawamura et al., Proc. of SRF2011, pp. 165–168, 2011
3. P. Kneisel et al., Proc. Of 2005 Particle Accelerator Conf., pp. 4012-4014, 2005.
4. G. Wu et al., Proc. of the 12th Workshop on RF superconductivity (SRF2005), pp. 600-603, 2005

Measurement of Cosmic-Ray Muon-Capture X-Rays and Application to Nuclear Material Detection

Y. Shimbara*

Cyclotron Radioisotope Center, Tohoku University,
Aoba-ku, Sendai, Miyagi 980-8578, Japan
* shimbara@cyric.tohoku.ac.jp

T. Matsuzaki

Nishina Center, RIKEN,
Wako, Saitama 351-0198, Japan

K. Abe, K. Haga, A. Homma, N. Kikukawa, D. Murooka, M. Nagashima,
Y. Nakamura, T. Ogura, T. Sakai, D. Sera and K. Tashiro

Graduate School of Science and Technology, Niigata University,
Nishi-ku, Niigata 950-2181, Japan

J. Goto

Radio Isotope Center, Niigata University,
Chuo-ku, Niigata-shi, Niigata 951-8510, Japan

J. Ito, S. Kato, T. Ohtsubo, S. Ohya, K. Takeda, Y. Tsurumaki, A. Yajima and
T. Yoshikawa

Department of Physics, Niigata University,
Nishi-ku, Niigata 950-2181, Japan

T. Itoh and M. Tanaka

Research Center for Nuclear Physics, Osaka University,
Ibaraki, Osaka 567-0047, Japan

S. Suzuki

National Institute of Radiological Sciences,
Inage-ku, Chiba 263-8555, Japan

K. Yanagi

Research Center for Electron Photon Science, Tohoku University,
Taihaku-ku, Sendai, Miyagi 982-0826, Japan

Muonic X-ray measurement by the use of cosmic muon has a potential to identify nuclear material in containers. We performed a feasibility study by using an iron target. Two plastic scintillators detected incoming cosmic-ray muons and a veto scintillator identified muons stopped in the target. Germanium detectors in coincidence with the scintillators measured muonic X-ray energies. We clearly observed muonic X-ray peaks in the photon spectrum, of which the energies were consistent with known muonic X-ray energies. By using the obtained spectrum, input parameters of the Monte-Carlo simulation were checked. The simulation for uranium target showed that this method is promising.

Keywords: Cosmic-ray; muonic atom; X-ray; Non-destructive analysis.

1. Introduction

Detecting nuclear material such as uranium (U) and plutonium (Pu) concealed in containers is a global issue for nuclear nonproliferation. Recently, cosmic muon has become attractive tools for non-destructive search owing to the high penetration ability. Several countries are developing detector systems for nuclear materials by applying technique of cosmic muon tomography, which measures deflection of passing muons owing to the multiple scattering[1,3]. However, this method has a difficulty in accurately determining atomic numbers.

While most cosmic muons penetrate Earth's surface, some of them stop in materials on the ground[4]. On being stopped in materials, a negative muon enters atomic orbit and cascade down to the ground state with emitting muonic X-rays. Since muon has 207 times larger mass than electron, muon is deeply bound and results in high energy X-rays. Therefore, muonic X-rays can be measured by detectors located outside materials. Muonic X-ray energy increases as a function of atomic number and there are many precise data in terms of muonic X-ray energies. Therefore, if we measure muonic X-ray energies, we can accurately identify the elements of substance within the materials. However, muonic X-rays originated from cosmic muons are rare compared to the other natural background. Therefore, there have been no report on the muonic X-ray measurement using cosmic muon. Rosenberg and Bernstein recently performed an estimation whether cosmic-ray muon-capture X-rays can be used for the detection of enriched U and Pu located in cargo containers[5]. Muonic X-rays emitted from such nuclear materials had energies exceeding 3 MeV, and such high-energy natural background γ-rays are rare. They noted a few muonic X-ray events coinciding with the incoming cosmic muon might provide a notice of detection of nuclear materials. Their estimation, however, showed that the background events produced by incoming cosmic muon would overlap with the muonic X-ray events, and make the measurement difficult. Therefore, they concluded that the cosmic-ray muon-capture X-ray measurement for the detection of nuclear materials was not realistic. However, whether muonic X-ray originated from cosmic muon can be measured or not has still been unclear. If we use not only counters to detect incoming muons but also a veto counter to identify stopped muons, and allow longer measurement time, muonic X-rays originated from cosmic muons may be measurable. This report aims to show the feasibility of muonic X-ray measurement with cosmic muons. In addition, we present a simulation in terms of muonic X-ray measurement for U, and discuss about nuclear material detection.

2. Experiment

The apparatus shown in Figure 1 was placed on the second floor of a five-story concrete building. Three plastic scintillators (S1, S2, S3) were placed from the top to the bottom in parallel. The iron (Fe) target with a volume of 2.6×10^3 cm^3 was placed on a plywood stage between the S2 and S3 scintillators. The S1 and S2 scintillators identified incoming cosmic muons, and the S3 scintillator was used as a veto counter to identify the stopped muons. Three germanium detector (Ge1, Ge2, Ge3) surrounded the target.

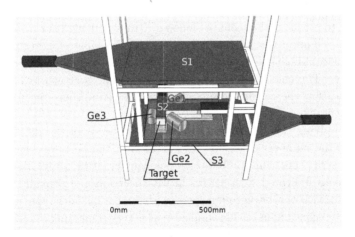

Fig. 1. Detector and target setup. The red cylinder indicates Ge1, with which the spectra were obtained.

The Ge1 detector was used to produce muonic X-ray spectra, and the Ge2 and Ge3 detectors were used as monitors. The efficiency and energy resolution of the Ge1 detector were 73 % relative to a 3-inch NaI(Tl) detector and 3.7 keV (FWHM) at 1.33 MeV, respectively. To effectively collect muonic X-rays with Ge detectors, the target was T-shaped, and the Ge detectors were placed so as to increase the detection solid angle as much as possible. The measurement times needed for clear identifications of muonic X-ray peaks was 16 days. The detail description about this experiment will be published somewhere soon[6].

3. Experimental result

The obtained spectra are shown in Figure 2. We clearly observed muonic X-ray peaks at 269 keV and 1255 keV in Fe. The obtained energies agreed with the known muonic X-ray energies[7] within the errors. This fact suggests that we can accurately determine the elements of the target. The large peak at 511 keV is electron-positron annihilation γ-ray, of which the main origin is thought of cosmic positron. A continuous background was observed, which extended to energies exceeding those of muonic X-rays. Existence of the background overlapping with the muonic X-ray peaks results in the long measurement time.

4. Simulation

In order to estimate the yields of muonic X-rays and the background from U target, we performed a simulation by using the Monte-Carlo simulation code of Geant4[8]. At first we checked the validity of the simulation on Fe target. The calculated muonic X-ray spectrum is shown in Fig. 2(b). We found that the calculated muonic X-ray spectrum reproduced well the experimental spectrum in Fig. 2(a).

By applying the same input parameters in terms of the detector system and the numbers of cosmic rays as used in the Fe target simulation, muonic X-ray spectrum for U target was calculated (See Fig. 3). Muonic X-ray peaks indicated with arrows in the spectrum had enough strengths compared with the background for the identification, although there was unexpected continuous background even at 6 MeV. Our simulation suggests that, if high efficiency X-ray detectors covering containers are applied, we may be able to identify muonic X-rays emitted from nuclear material within a few weeks.

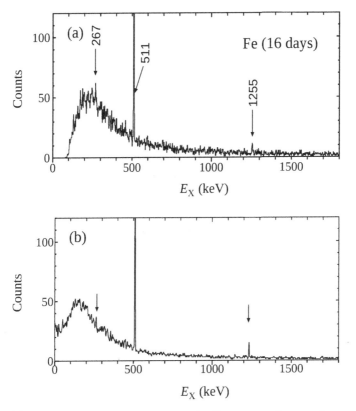

Fig. 2. (a) Measured muonic X-ray spectrum for Fe target, and comparison with (b) spectrum produced by Monte-Carlo simulation code of Geant4. The arrows in (a) indicate the energies of the muonic X-ray peaks, whereas the peak at 511 keV is originated from electron-positron annihilation γ-rays. The arrows in (b) indicates the corresponding muonic X-ray peaks.

5. Summary

We studied the feasibility of cosmic-ray muon capture X-ray measurement. Employing two scintillators to detect incoming cosmic muons, and one VETO scintillator to identify muons stopped in the target, the Ge detectors measured muonic X-rays. Measurement time required to identify muonic X-rays was 16 days. We clearly observed muonic X-ray peaks in the photon spectrum. This fact suggests that we can accurately determine elements of substance by measuring muonic X-rays originated from cosmic muons. In addition, we performed the Monte-Carlo simulation for U target. We found that the muonic X-ray peaks were detectable with the present setup. If we

338

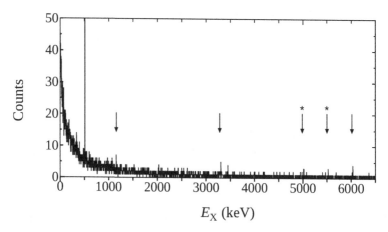

Fig. 3. Muonic X-ray spectrum for U target calculated with Monte-Carlo simulation code of Geant4. Arrows indicate muonic X-ray peaks, where asterisks indicate the escape peaks.

apply high efficiency muonic X-ray detectors covering containers, we may be able to identify muonic X-ray emissions from the inner nuclear materials in a few weeks.

References

1. K. N. Borozdin, G. E. Hogan, C. Morris, W. C. Priedhorsky, A. Saunders, L. J. Schultz, and M. E. Teasdale, Radiographic imaging with cosmic-ray muons, *Nature* **422** (2003) 277.
2. C. L. Morris, C. C. Alexander, J. D. Bacon, K. N. Borozdin, D. J. Clark, R. Chartrand, C. J. Espinoza, A. M. Fraser, M. C. Galassi, J. A. Green, J. S. Gonzales, J. J. Gomez, N. W. Hengartner, G. E. Hogan, A. V. Klimenko, M. F. Makela, P. McGaughey, J. J. Medina, F. E. Pazuchanics, W. C. Priedhorsky, J. C. Ramsey, A. Saunders, R. C. Schirato, L. J. Schultz, M. J. Sossong, and G. S. Blanpied, Tomographic imaging with cosmic ray muons, *Sci. Glob. Secur.* **16** (2008) 37.
3. S. Pesente, S. Vanini, M. Benettoni, G. Bonomi, P. Calvini, P. Checchia, E. Conti, F. Gonella, G. Nebbia, S. Squarcia, G. Viesti, A. Zenoni, G. Zumerle, First results on material identification and imaging with a large-volume muon tomography prototype, *Nucl. Inst. Meth. A* **604** (2009) 738.
4. P.K.F. Grieder, *Cosmic rays at earth : researcher's reference manual and data book* (ElsevierScience, Amsterdam, 2001) pp. 459-668.

5. L. Rosenberg, and A. Bernstein, *Feasibility of Sea-level Cosmic-Ray Muon-Capture SNM Detection*, Lawrence Livermore Natl. Lab. UCRL-TR-210517 (2005).

6. Y. Shimbara, T. Matsuzaki, K. Abe, J. Goto, T. Itoh, N. Kikukawa, M. Nagashima, Y. Nakamura, T. Ogura, T. Ohtsubo, S. Ohya, T. Sakai, D. Sera, S. Suzuki, K. Takeda, A. Yajima, T. Yoshikawa, to be published.

7. D.F. Measday, The nuclear physics of muon capture, *Phys. Rep.* **354** (2001) 243.

8. S. Agostinelli et al., Geant4 - a simulation toolkit, *Nucl. Inst. Meth. A* **506** (2003) 250, http://geant4.cern.ch/.

Active Neutron-Based Interrogation System with D-D Neutron Source for Detection of Special Nuclear Materials[*]

Y. Takahashi[1†], T. Misawa[1], T. Yagi[1], C. H. Pyeon[1], M. Kimura[1], K. Masuda[2] and H. Ohgaki[2]

[1]*Research Reactor Institute, Kyoto University*
Asashiro-Nishi, Kumatori-cho, Sennan-gun, Osaka, 590-0494, Japan
[2]*Institute of Advanced Energy, Kyoto University*
Gokasho, Uji, Kyoto 611-0011, Japan
[†] *ytaka@rri.kyoto-u.ac.jp*

The detection of special nuclear materials (SNM) is an important issue for nuclear security. The interrogation systems used in a sea port and an airport are developed in the world. The active neutron-based interrogation system is the one of the candidates. We are developing the active neutron-based interrogation system with a D-D fusion neutron source for the nuclear security application. The D-D neutron source is a compact discharge-type fusion neutron source called IEC (Inertial-Electrostatic Confinement fusion) device which provides 2.45 MeV neutrons. The nuclear materials emit the high-energy neutrons by fission reaction. High-energy neutrons with energies over 2.45 MeV amount to 30% of all the fission neutrons. By using the D-D neutron source, the detection of SNMs is considered to be possible with the attention of fast neutrons if there is over 2.45 MeV. Ideally, neutrons at En>2.45 MeV do not exist if there is no nuclear materials. The detection of fission neutrons over 2.45 MeV are hopeful prospect for the detection of SNM with a high S/N ratio. In the future, the experiments combined with nuclear materials and a D-D neutron source will be conducted. Furthermore, the interrogation system will be numerically investigated by using nuclear materials, a D-D neutron source, and a steel container.

Keywords: Active interrogation of special nuclear materials; counter terrorism; D-D neutron source; organic scintillator, fission neutron; TENA.

1. Introduction

The detection of special nuclear materials (SNM) hidden in containers is urgent and crucial concern in the world. Therefore, thedevelopment of an interrogation system for nuclear security is strongly demanded. Various techniques are proposed as interrogation systems for detection of SNM. The active neutron-

---

[*] This work is supported by the R&D Program for Implementation of Anti-Crime and Anti-Terrorism Technologies for a Safe and Secure Society promoted by Japan Science and Technology Agency.

based interrogation system is the one of the candidates (see Fig. 1) [1, 2]. The R&D project of the interrogation system started since 2010 [3, 4]. We are developing the active neutron-based interrogation system with a D-D neutron source. The D-D neutron source is a compact discharge-type fusion neutron source called IEC (Inertial-Electrostatic Confinement fusion) developed by Kyoto University [5-9], which provides 2.45 MeV fusion neutrons. The advantage of the D-D neutron source for practical use are as follows; safety, easy maintenance, without the need of tritium handling, less shielding than D-T source, robust, compact, possible to operate pulse mode and CW mode and so on. To take advantage of the D-D neutron source, we have developed two advanced active neutron-based method, namely delayed neutron noise analysis method (DNNA) and threshold energy neutron analysis method (TENA). Since the detail description of the DNNA method is given in this proceedings [10], the TENA method is discussed in this paper.

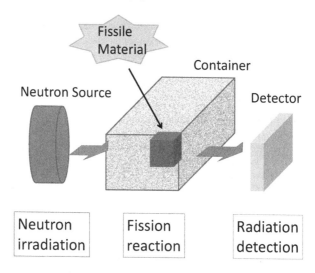

Fig. 1 Concept of the active neutron-base interrogation system.

2. Threshold Energy Neutron Analysis (TENA) Method

A D-D neutron source emits 2.45 MeV neutrons by the fusion reaction. On the other hands, nuclear materials emit high-energy neutrons by the fission reaction. High-energy neutrons with energies over 2.45 MeV amount to 30% in the fission neutrons as shown in Fig.2. If there is no nuclear materials, neutrons over 2.45 MeV do not ideally exist. It means the method can obtain the high S/N ratio.

This method has a great advantage; it is possible for us to useboth a pulsed D-D neutronsource and CW D-D neutron source. In addition, this method can be used in combination with other methods. Therefore, it is possible to improve the detection accuracy of the interrogation system.

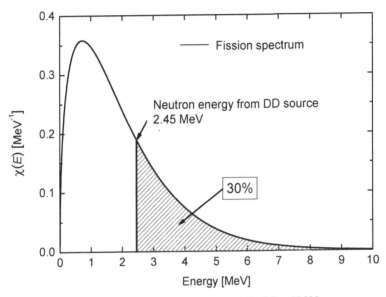

Fig. 2 Energy spectrum of fission neutrons emitted from U-235.

3. Experiment with D-D Neutron Source

For the detection of fission neutrons, a liquid organic scintillation detector was used in this study as shown in Fig. 3. It contains hydrogen atoms, which are recoiled by collision with fast neutron and produce scintillation light. Gamma-rays entered into the detectors also produce scintillation light. Neutron and gamma-ray events can identified by using the rise-time discrimination technique as shown in Fig. 4. The pulse height distribution of discriminated signals caused by neutron are related with neutron energy spectrum.

344

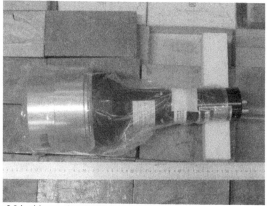

Fig. 3 Liquid organic scintillation detector with a photo multiplier tube.

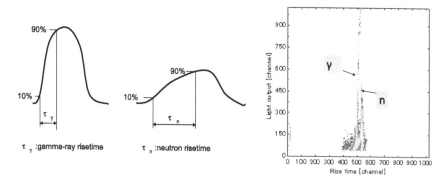

Fig. 4 Rise-time discrimination technique.
(Left: signal from preamp., Right: 2 dimensional pulse height distribution)

First of all, to demonstrate the validity of the method, the experiment was carried out with a D-D neutron source and a Cf-252 neutron source as shown in Fig, 5. We measured the neutron spectrum in three different cases, namely, (1) with only the D-D neutron source, (2) with only the Cf-252 neutron source, and (3) together withthe D-D neutron source and the Cf-252 neutron source. The Cf-252 neutron source was used for simulated fissile materials because the neutron energy spectrum obtained with Cf-252 is similar to that of fission reaction of U-235. The results of measured pulse height distributions of the liquid organic scintillator for neutron detection is shown in Fig.6, which demonstrates that the corresponding signals by neutrons whose energy is higher than 2.45 MeV was detected even without Cf-252 (case 1). However, neutrons from the Cf-252 can be measured even with the D-D operation conditions (case 3). This result indicates nuclear materials can be detected by the method and the detectorsystem.

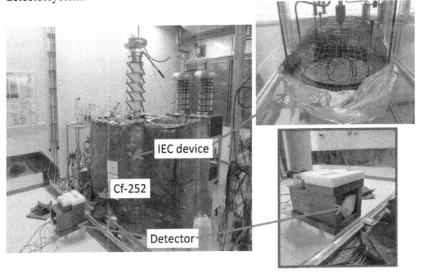

Fig. 5 Experimental setup with the D-D neutron source.

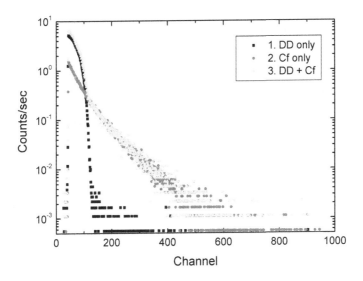

Fig. 6 Measured pulse height distributions of liquid organic scintillator for neutron detection.

4. Concluding Summary

We can measure neutrons from Cf-252 which is a simulated material for U-235 even with IEC device operation. It indicates this method can detected a SNM. For more precise detection by this method, it is required to reduce noise signals during D-D neutron source operation by using appropriate shielding materials for detector or development of signal process system.

References

1. R. F. Radel et al., *Fusion Sci. and Technol.* 52, 1087 (2007) .
2. K. A. Jordan et al., Nucl. Instrum. Meth. A 579, 388 (2007).
3. H. Ohgaki et al., Conceptual Design of a Nuclear Material Detection System Based on the Neutron / Gamma-Ray Hybrid Approach, in *Proc. of 2010 IEEE International Conference on Technologies for Homeland Security (HST 2010)*, (Waltham, MA, USA, 2010) pp.525-529.
4. H. Ohgaki et al., Non-Destructive Inspection System for Special Nuclear Material using Inertial Electrostatic Confinement Fusion Neutrons and Laser Compton Scattering Gamma-rays, in *Proc. of 2012 IEEE International Conference on Technologies for Homeland Security (HST 2012)*, (Waltham, MA, USA, 2012) pp. 666-671.
5. K. Yoshikawa et al., Research and Development of the Humanitarian Landmine Detection System by a Compact Fusion Neutron Source, *IEEE Trans. Nuclear Sci.* 56-3, 1193 (2009).
6. K. Masuda et al., Compact Intense Neutron Generators based on Inertial Electrostatics Confinement of D-D fusion Plasmas, in this proceedings.
7. T. Takamatsu et al., Spatial Distribution of D-D Neutrons of a Compact Water-Cooled Inertial Electrostatic Confinement Device, *Fusion Sci. Tech.* 52-4, 1114 (2007).
8. K. Masuda et al., Research and Development of Compact Neutron Sources based on Inertial Electrostatic Confinement Fusion, in *Proc. of 20th International Conference on the Application of Accelerators in Research and Industry (CAARI 2008)*, (Fort Worth, TX, USA, 2008) pp.587-590.
9. K. Masuda et al., Numerical Study of Ion Recirculation in an Improved Spherical Inertial Electrostatic Confinement Fusion Scheme by Use of a Multistage High-Voltage Feedthrough, *Fusion Sci. Tech.* 60-2, 625 (2011).
10. T. Misawa et al., Development of Measurement Methods for Detection of Special Nuclear Materials using D-D Pulsed Neutron Source, in this proceedings.

Closing Summary

Summary Comments: Nuclear Physics and Gamma-Ray Sources for Nuclear Security and Nonproliferation

C. P. J. Barty

Lawrence Livermore National Laboratory,
Livermore, CA 94551, USA
barty1@llnl.gov

The Nuclear Physics and Gamma-ray Sources for Nuclear Security and Nonproliferation (NPNSNP) meeting held in Tokai-mura, Japan from January 28th to 30th, 2014 revealed both the rapid evolution and growth of mono-energetic, laser-Compton, gamma-ray source technology and the emergence of numerous important applications enabled by this technology. More than $500M of large-scale source and development activities were represented at the meeting, including all of the major projects in the United States, Europe and Japan. The meeting was both highly stimulating intellectually and provided an excellent venue for the exploration of new collaborations between groups.

The Tokai meeting demonstrated though its many presentations that the next generation of gamma-ray sources currently under development and the next generation of detection technologies that utilize these sources will meet the needs of the nuclear materials management community for important missions such as quantitative characterization of nuclear materials removed from Fukushima. Many related and important societal applications of next generation gamma-ray sources, e.g. spent fuel management, clandestine material identification etc. were also discussed and debated at the Tokai meeting. In this regard the meeting clearly addressed its namesake charter and was a resounding success.

An important aspect of the Tokai meeting was the fact that the majority of the presentations were made by nuclear physicists and chemists that were interested in "using' new, mono-energetic gamma-ray capabilities and not in "developing" them. This transition from machine science to applications science marks an important transition and expansion for the "nuclear photonics" community and suggests that future meetings will have a wide appeal beyond the primary nuclear safeguards application area emphasized at Tokai. At the conclusion of the NPNSNP conference it was agreed that the next meeting should be held in 2016 and that this next meeting should be expanded to become the first International Nuclear Photonics Conference. This new meeting will emphasize the unique ability of mono-energetic gamma-ray sources to solve problems facing the nuclear non-proliferation community as well as welcome the discussion and presentation of other gamma-ray-enabled applications such as precision photo-nuclear spectroscopy, isotope-specific medical radiography and radiology, industrial nondestructive evaluation, nuclear fuel management and waste characterization, etc.

Acknowledgment

This work was performed under the auspices of the U.S. Department of Energy by the Lawrence Livermore National Laboratory under Contract DE-AC52-07NA27344.

Program Schedule of the International Workshop

28 January (Tuesday)
9:00-9:15 Coffee

Session 1 (Chair: Yosuke Naoi)
9:15-9:20 Opening Remarks/Shojiro Matsuura (President of JAEA, Japan)
9:20-9:30 Welcome Speech/Paul Bolton (QuBS, JAEA, Japan)
9:30-10:00 Speech/Nuclear Security Activity of JAEA/Masao Senzaki (ISCN, JAEA, Japan)

10:00-10:20 Group Photo

Session 2 (Chair: Calvin R. Howell)
10:20-10:50 Nuclear Science and Applications with next generation of High Power Lasers and Brilliant Low Energy Gamma Beams at ELI-NP/Sydney Gales (ELI-NP, Romania)
10:50-11:20 New Non-intrusive inspection technologies for nuclear security and nonproliferation/Robert J. Ledoux (Passport Systems Inc., USA)
11:20-11:50 Development of gamma-ray NDA for Nuclear Safeguards and Security at JAEA/Ryoichi Hajima (JAEA, Japan)

11:50-13:20 Lunch and Poster Session

Session 3 (Chair: Toshiyuki Shizuma)
13:20-13:50 Photonuclear Reaction Studies at HI γS: Developing the Science of Remote Detection of Nuclear Materials/Calvin R. Howell (Duke Univ./TUNL, USA)
13:50-14:20 Dipole strength around the particle threshold/Deniz Savran(EMMI/GSI, Germany)
14:20-14:40 Resonant photonuclear excitations for neutrino nuclear responses and nuclear isotope detections/Hiroyasu Ejiri (Osaka Univ., Japan)

14:40-15:00 Coffee

Session 4 (Chair: Hiroyasu Ejiri)
15:00-15:30 Nuclear Resonance Fluorescence in Pu-240 and Spent Nuclear Fuel Assay/Bernhard A. Ludewigt (LBNL, USA)
15:30-15:50 Non-Destructive Testing for Control of Radioactive Waste Package/Stéphane Plumeri (National Radioactive Waste Management Agency, France)
15:50-16:10 Development of first responders' equipment at RN incident sites/Ken'ichi Tsuchiya (National Research Institute of Police Science, Japan)

16:10-16:30 Coffee

Session 5 (Chair: Vladimir G. Nedorezov)
16:30-17:00 SCAPA - the Scottish Centre for the Application of Plasma-based Accelerators/Dino Jaroszynski (University of Strathclyde, UK)
17:00-17:30 Compact Short-Pulsed Electron Linac Based Neutron Sources for Precise Nuclear Material Analysis/Mitsuru Uesaka (University of Tokyo, Japan)
17:30-18:00 Compact quasi-monoenergetic gamma ray sources based on Laser Plasma Accelerators/Cameron G.R. Geddes (LBNL, USA)
18:00-18:30 The Overview of Optical Cavity R&D for γ-ray generation based on Inverse Compton Scattering/Junji Urakawa (KEK, Japan)

18:30-20:00 Dinner

29 January (Wednesday)
Session 6 (Chair: Hideaki Ohgaki)
9:00-9:30 Advanced laser-Compton gamma-ray sources for nuclear materials detection, assay and imaging/Christopher P. J. Barty (LLNL, USA)
9:30-9:50 Compact Gamma-Beam Source Based on Compton Storage Ring for Nuclear Security/Peter Gladkikh (NSC, KIPT, Kharkov, Ukraine)
9:50-10:10 Developments of optical resonators and optical recirculators for Compton X/gamma ray machines/Aurelien Martens (LAL/IN2P3, France)

10:10-10:40 Coffee

Session 7 (Chair: Shigeru Kubono)
10:40-11:00 Low-lying "Pygmy" Dipole Resonances and Strength Functions/Volker Werner (Technical University of Darmstadt, Germany)
11:00-11:20 NRF-Based NDA of Nuclear Material using Monochromatic γ-ray Beam/Toshiyuki Shizuma (JAEA, Japan)
11:20-11:40 Improving the Assay of 239Pu in Spent and Melted Fuel Using the

Nuclear Resonance Fluorescence Integral Resonance Transmission Method/Christopher T. Angell (JAEA, Japan)

11:40-13:00 Lunch

Session 8 (Chair: Sydney Gales)
13:00-13:30 Laser Compton Scattering Gamma-ray Beam Source at NewSUBARU Storage Ring/Shuji Miyamoto (University of Hyogo, Japan)
13:30-14:00 Energy Calibration of Electron and Gamma-ray Beams at NewSUBARU-GACKO/Tatsushi Shima (Osaka Univ., Japan)
14:00-14:20 A Paradigm for the nondestructive assay of spent fuel assemblies and similar large objects, with emphasis on the role of photon-based techniques/Alan bolind (JAEA, Japan)
14:20-14:40 Medium effect in the photoexcitation of α cluster in 6Li/Tamio Yamagata (Konan Univ., Japan)

14:40-15:00 Coffee

Session 9 (Chair: Michio Seya)
15:00-15:30 Nuclear Safeguard with Cosmic-Ray-Muon Imaging/Haruo Miyadera (Toshiba Corporation, Japan)
15:30-15:50 Compact radiation for nuclear waste management and transmutation/Eugene Bulyak (NSC KIPT, Kharkov, Ukraine)
15:50-16:10 Compact Intense Neutron Generators based on Inertial Electrostatic Confinement of D-D Fusion Plasmas/Kai Masuda (Kyoto Univ., Japan)

16:10-16:40 Coffee

Session 10 (Chair: Bernhard A. Ludewigt)
16:40-17:10 NRF Based Nondestructive Inspection System for SNM by Using Laser-Compton-Backscattering Gamma-rays/Hideaki Ohgaki (IAE, Kyoto Univ., Japan)
17:10-17:40 Development of Measurement Methods for Detection of Special Nuclear Materials using D-D Pulsed Neutron Source/Tsuyoshi Misawa (RRI, Kyoto Univ., Japan)
17:40-18:20 Challenges of a new proposed Compton Source for the ELI-NP European Infrastructure/Luca Serafini (INFN-Milano, Italy)

30 January (Thursday)

Session 11 (Chair: Yoshiaki Kiyanagi)

9:00-9:30 SOFIA, a next-generation facility for fission yields measurements and fission study: First results and perspectives/Laurent Audouin (IPNOrsay, University of Paris Sud, France)

9:30-10:00 Present Status of Nuclear Data for Nuclear Nonproliferation/Masayuki Igashira (Tokyo Institute of Technology, Japan)

10:00-10:20 Development of the experimental photo-nuclear reaction Database in Hokkaido University Nuclear Data Center/Ayano Makinaga (Hokkaido Univ., Japan)

10:20-10:40 Coffee

Session 12 (Chair: Dino Jaroszynski)

10:40-11:10 Laser Compton scattering photon beams and other gamma-ray sources: Project for coherent γ-ray source on basis of femtosecond laser at ILC MSU/Vladimir G. Nedorezov (RAS, INR, Russia)

11:10-11:30 Laser driven ion acceleration study in JAEA/Kiminori Kondo (JAEA, Japan)

11:30-11:50 Status of new JENDL Photonuclear Data File/Kazuaki Kosako (Shimizu Corporation, Japan)

11:50-12:50 Lunch

Session 13 (Chair: Mamoru Fujiwara)

12:50-13:20 Prospect of Nuclear Research at a laser-photon Storage ring electron Compton Scatter Beamline of 15 MeV γ- Rays/Chary Rangacharyulu (University of Saskatchewan, Canada)

13:20-13:50 IRIDE: An Interdisciplinary Research Infrastructure based on Dual Electron linac & laser/Massimo Ferrario (INFN-LNF, Italy)

13:40-14:00 Closing Remarks/Christopher P. J. Barty (LLNL, USA)

14:00-14:10 Final Remarks/Mamoru Fujiwara (Osaka Univ., Japan)

14:20-18:30 Excursion to KEK Compact ERL(Return to Riccoti)

[Poster Presenter]

A Study of the Nuclear Resonance Fluorescence Reaction Yield Dependence on the Target Thickness of ^{208}PB/Hani Negm

Photodisintegration reactions with linear polarized gamma-ray beam/Hayakawa Takehito (JAEA, Japan)

Test Experiment of γ-ray Diffraction for Crystal Monochromators/Shunya Matsuba (JAEA, Japan)

Overview of laser Compton-scattered photon source at the cERL/Ryoji
Nagai (JAEA, Japan)
Development of a high-brightness and high-current electron gun
for high-flux γ-ray generation/Nobuyuki Nishimori (JAEA, Japan)
DESIGN OF ERL SPOKE CAVITY FOR NON-DESTRUCTIVE ASSAY
RESEARCH /Masaru Sawamura (JAEA, Japan)
Measurement of cosmic-ray muon-capture X-rays and its application to nuclear
material detection/Yoshihiro Shimbara (Tohoku Univ., Japan)
Development of Active Neutron-based Interrogation System with D-D neutron
source for Detection of Special Nuclear Materials/Yoshiyuki Takahashi (Kyoto
Univ., Japan)

List of Participants

Masao Senzaki
Integrated Support Center for Nuclear Nonproliferation and Nuclear Security
(ISCN), Japan Atomic Energy Agency (JAEA), 3-1-1 Funaishikawaekihigashi,
Tokai-mura, Naka-gun, Ibaraki 319-1118, Japan
senzaki.masao@jaea.go.jp

Sydney Gales
Scientific Director, Extreme Light Infrastructure-Nuclear Physics (ELI-NP)/
Horia Hulubei National Institute of Physics and Nuclear Engineering (IFIN-HH)
30 reactorului st, PO BOX MG-6, Bucharest Magurele, 077125, Romania
gales@in2p3.fr
sydney.gales@eli-np.ro

Robert J. Ledoux
President/CEO, Passport Systems, Inc., 70 Treble Cove Road, North Billerica,
MA 01862, USA
ledoux@passportsystems.com

Ryoichi Hajima
Quantum Beam Science Directorate (QUBS), Japan Atomic Energy Agency
(JAEA), 2-4 Shirane, Shirakata, Tokai-mura, Naka-gun, Ibaraki 319-1195, Japan
hajima.ryoichi@jaea.go.jp

Calvin R. Howell
Professor and Director, Physics Department of Duke University and Triangle
Universities Nuclear Laboratory (TUNL), Box 90308, Durham, NC 27708, USA
howell@tunl.duke.edu

Deniz Savran
ExtreMe Matter Institute EMMI and Research Division,
GSI Helmholtzzentrum für Schwerionenforschung GmbH, Darmstadt, Germany
and Frankfurt Institute for Advanced Studies FIAS, Frankfurt am Main,
Germany
d.savran@gsi.de

Hiroyasu Ejiri
Research Center for Nuclear Physics (RCNP), Osaka University, Mihogaoka, Ibaraki, Osaka, Japan
ejiri@rcnp.osaka-u.ac.jp

Bernhard A. Ludewigt
Scientist, Accelerator and Fusion Research Division (AFRD), Lawrence Berkeley National Laboratory (LBNL), 1 Cyclotron Road, MS 50A2177, Berkeley, CA 94707, USA
baludewigt@lbl.gov

Stéphane Plumeri
Research & Development, National radioactive waste management agency (ANDRA), 1-7, rue Jean-Monnet - 92298 Chatenay-Malabry cedex, France
stephane.plumeri@andra.fr

Ken'ichi Tsuchiya
Researcher, National Research Institute of Police Science (NRIPS), 6-3-1 Kashiwanoha, Kashiwa, Chiba 277-0882, Japan
tsuchiya@nrips.go.jp

Dino Jaroszynski
Physics Department, University of Strathclyde, John Anderson Building, 107 Rottenrow, Glasgow, G4 0NG, UK
dino@phys.strath.ac.uk

Mitsuru Uesaka
Professor, Nuclear Professional School, University of Tokyo, 22-2 Shirane-shirakata, Tokai, Naka, Ibaraki, Japan
uesaka@nuclear.jp

Cameron G.R. Geddes
The Lasers, Optical Accelerator Systems Integrated Studies (LOASIS), Lawrence Berkeley National Laboratory (LBNL), 1 Cyclotron Road, MS71-259, USA
cgrgeddes@lbl.gov

Junji Urakawa
Professor, Accelerator, High Energy Accelerator Research Organization (KEK), 1-1 Oho Tsukuba-shi, Ibaraki, Japan
junji.urakawa@kek.jp

Christophe P. J. Barty
Chief Technology Officer, National Ignition Facility & Photon Science
Directorate (NIF&PS), Lawrence Livermore National Laboratory (LLNL),
Livermore, CA 94551 USA
barty1@llnl.gov

Peter Gladkikh
Cyclic accelerators, National Science Center Kharkov Institute of Physics and
Technology (NSC KIPT), 1 Academicheskaya St., 61108, Kharkov, Ukraine
gladkikh@kipt.kharkov.ua

Aurelien Martens
Postdoctoral, Accelerator Department, Laboratoire de l'Accélérateur Linéaire
(LAL) Institut national de physique nucléaire et de physique des particules
(IN2P3)/ Centre national de la recherche scientifique (CNRS), Centre Scienfique
d'Orsay, Bat. 200 ; BP 34 ; 91898 ORSAY Cedex, France
martens@lal.in2p3.fr

Volker Werner
Research Scientist, Institute for Nuclear Physics, Technical University
Darmstadt, Schlossgartenstrasse 9, 64289 Darmstadt, Germany
volker.werner@yale.edu

Toshiyuki Shizuma
Quantum Beam Science Directorate (QUBS), Japan Atomic Energy Agency
(JAEA), 2-4 Shirane, Shirakata, Tokai-mura, Naka-gun, Ibaraki 319-1195, Japan
shizuma.toshiyuki@jaea.go.jp

Christophe T. Angell
Quantum Beam Science Directorate (QUBS), Japan Atomic Energy Agency
(JAEA), 2-4 Shirane, Shirakata, Tokai-mura, Naka-gun, Ibaraki 319-1195, Japan
angell.christopher@jaea.go.jp

Shuji Miyamoto
Director, Laboratory of Advanced Science and Technology for Industry,
University of Hyogo, 1-1-2 Kouto, Kamigori-cho, Ako-gun, Hyogo, 678-1205,
Japan
miyamoto@lasti.u-hyogo.ac.jp

Tatsushi Shima
Assistant Professor, Research Center for Nuclear Physics, Osaka University,
10-1, Mihogaoka, Ibaraki, Osaka, Japan
shima@rcnp.osaka-u.ac.jp

Alan Bolind
Integrated Support Center for Nuclear Nonproliferation and Nuclear Security
(ISCN), Japan Atomic Energy Agency (JAEA), 3-1-1 Funaishikawaekihigashi,
Tokai-mura, Naka-gun, Ibaraki 319-1118, Japan
alan.bolind@jaea.go.jp

Tamio Yamagata
Professor, Department of Physics, Konan University, Okamoto 8-9-1,
Higashinada, Kobe 658-8501, Japan
yamagata@center.konan-u.ac.jp

Haruo Miyadera
Scientist, Power System Company, Toshiba Corporation, 8, Shinsugita-Cho,
Isogo-Ku, Yokohama, Kanagawa 235-8523, Japan
haruo.miyadera@toshiba.co.jp

Eugene Bulyak
Senior researcher, physicist, Institute of High Energy and Nuclear Physics,
National Science Center "Kharkov Institute of Physics and Technology" (NSC
KIPT), 1 Academicheskaya St., 61108, Kharkov, Ukraine
bulyak@kipt.kharkov.ua

Kai Masuda
Associate professor, Institute of Advanced Energy, Kyoto University, Gokasho,
Uji, Kyoto 611-0011, Japan
masuda@iae.kyoto-u.ac.jp

Hideaki Ohgaki
Professor, Institute of Advanced Energy (IAE), Kyoto University, Gokasho, Uji,
Kyoto 611-0011 JAPAN
ohgaki@iae.kyoto-u.ac.jp

Tsuyoshi Misawa
Professor, Department of Nuclear Science and Engineering, Research Reactor
Institute (RRI), Kyoto University, 2-1010 Asashiro-nishi, Kumatori, Osaka 590-
0494 JAPAN
misa@rri.kyoto-u.ac.jp

Luca Serafini
Research Director, Istituto Nazionale di Fisica Nucleare (INFN) , Sezione di
Milano, Via Celoria 16 - 20133, Milano - Italy
luca.serafini@mi.infn.it

Laurent Audouin
Lecturer & Researcher, Institut de Physique Nucléaire d'Orsay, Université Paris-Sud, Batiment 100, avenue Georges Clémenceau, 91406 Orsay cedex, France
audouin@ipno.in2p3.fr

Masayuki Igashira
Professor, Research Laboratory for Nuclear Reactors (RLNR), Tokyo Institute of Technology, 2-12-1-N1-26 Ookayama, Meguro-ku, Tokyo 152-8550, Japan
iga@nr.titech.ac.jp

Ayano Makinaga
Assistant professor, Faculty of Science, Hokkaido University, Nuclear Database Development Center, Main building N211, N10, W8, Kita-ku, Sapporo, 060-0810, Hokkaido, Japan
Makinaga@nucl.sci.hokudai.ac.jp

Vladimir G. Nedorezov
Head of Photo nuclear Laboratory, Institute for Nuclear Research (INR), Russian Academy of Sciences (RAS), Institute for Nuclear Research, 117312 Moscow, prospect 60-let Oktabra, 7A Russia
vladimir@cpc.inr.ac.ru

Kiminori Kondo
Unit Manager, Advance Beam Research Division (QUBS), Quantum Beam Science Directorate, Japan Atomic Energy Agency (JAEA), 8-1-7 Umemidai, Kizugawa-shi, Kyoto, 619-0215, Japan
kondo.kiminori@jaea.go.jp

Kazuaki Kosako
Researcher, Institute of Technology, Center for Atomic Energy Engineering, Shimizu Corporation, 4-17, Etchujima 3-Chome, Koto-Ku, Tokyo 135-8530, Japan
k.kosako@shimz.co.jp

Chary Rangacharyulu
Professor and Head, Physics and Engineering Physics, 116- Science Place, University of Saskatchewan, Saskatoon, SK, Canada, S7N 5E2
chary.r@usask.ca

Massimo Ferrario
Senior Scientist, Accelerator Division, Istituto Nazionale di Fisica Nucleare-
Laboratori Nazionali di Frascati (INFN-LNF), Vie E. Fermi 40, 00044 Frascati
(Roma), Italy
massimo.ferrario@lnf.infn.it

Hani Negm
PhD. Student, Quantum Radiation Energy, Institute of Advanced Energy, Kyoto
University, Gokasho, Uji-shi, Kyoto-fu 611-0011, Japan
haninegm@iae.kyoto-u.ac.jp

Hayakawa
Takehito
Quantum Beam Science Directorate (QUBS), Japan Atomic Energy Agency
(JAEA), 2-4 Shirane, Shirakata, Tokai-mura, Naka-gun, Ibaraki 319-1195, Japan
hayakawa.takehito@jaea.go.jp

Shunya Matsuba
Quantum Beam Science Directorate (QUBS), Japan Atomic Energy
Agency(JAEA), 2-4 Shirane, Shirakata, Tokai-mura, Naka-gun, Ibaraki 319-
1195, Japan
matsuba.shunya@jaea.go.jp

Ryoji Nagai
Quantum Beam Science Directorate (QUBS), Japan Atomic Energy
Agency(JAEA), 2-4 Shirane, Shirakata, Tokai-mura, Naka-gun, Ibaraki 319-
1195, Japan
nagai.ryoji@jaea.go.jp

Nobuyuki Nishimori
Quantum Beam Science Directorate (QUBS), Japan Atomic Energy Agency
(JAEA), 2-4 Shirane, Shirakata, Tokai-mura, Naka-gun, Ibaraki 319-1195, Japan
nishimori.nobuyuki@jaea.go.jp

Masaru Sawamura
Quantum Beam Science Directorate (QUBS), Japan Atomic Energy Agency
(JAEA), 2-4 Shirane, Shirakata, Tokai-mura, Naka-gun, Ibaraki 319-1195, Japan
sawamura.masaru@jaea.go.jp

Yoshihiro Shimbara
Research associate, Cyclotron and Radioisotope Center (CYRIC), Tohoku
University, 6-3 Aoba, Aramaki, Aoba-ku, Sendai, Miyagi, 980-8578 Japan
shimbara@cyric.tohoku.ac.jp

Yoshiyuki Takahashi
Nuclear Engineering Science Division, Research Reactor Institute, Kyoto
University, Kumatori-cho, Sennan-gun, Osaka 590-0494 Japan
ytaka@rri.kyoto-u.ac.jp

Kenro Kuroki
National Research Insitute of Police Science(NRIPS), 6-3-1 Kashiwanoha,
Kashiwa, Chiba 277-0882, Japan

Leontina A. Banu
Assistant Professor, Physics and Astronomy, James Madison University, 2031
Willow Hill Drive, Harrisonburg, VA 22801
banula@jmu.edu

Michio Seya
Integrated Support Center for Nuclear Nonproliferation and Nuclear Security
(ISCN), Japan Atomic Energy Agency (JAEA), 3-1-1 Funaishikawaekihigashi,
Tokai-mura, Naka-gun, Ibaraki 319-1118, Japan

Mai Than
International Cooperation Division, Vietnam Agency for Radiation and Nuclear
Safety(VARANS)

Yosuke Naoi
Integrated Support Center for Nuclear Nonproliferation and Nuclear Security
(ISCN), Japan Atomic Energy Agency (JAEA), 3-1-1 Funaishikawaekihigashi,
Tokai-mura, Naka-gun, Ibaraki 319-1118, Japan

Naoki Kobayashi
Integrated Support Center for Nuclear Nonproliferation and Nuclear Security
(ISCN), Japan Atomic Energy Agency (JAEA), 3-1-1 Funaishikawaekihigashi,
Tokai-mura, Naka-gun, Ibaraki 319-1118, Japan

Shigeru Kubono
Nishina Center, RIKEN, 2-1, Hirosawa, Wako, Saitama 351-0198

Kazuhisa Kakurai
Quantum Beam Science Directorate (QUBS), Japan Atomic Energy Agency
(JAEA), 2-4 Shirane, Shirakata, Tokai-mura, Naka-gun, Ibaraki 319-1195, Japan

Wataru Utsumi
Quantum Beam Science Directorate (QUBS), Japan Atomic Energy Agency
(JAEA), 2-4 Shirane, Shirakata, Tokai-mura, Naka-gun, Ibaraki 319-1195, Japan

Paul Bolton
Quantum Beam Science Directorate (QUBS), Japan Atomic Energy Agency (JAEA), 2-4 Shirane, Shirakata, Tokai-mura, Naka-gun, Ibaraki 319-1195, Japan

Akira Sugiyama
Quantum Beam Science Directorate (QUBS), Japan Atomic Energy Agency (JAEA), 2-4 Shirane, Shirakata, Tokai-mura, Naka-gun, Ibaraki 319-1195, Japan

Yoshihiro Nakagome
President, Japan Nuclear Energy Safety Organization (JNES), 1-9-9, Roppongi, Minato-ku, Tokyo, Japan 106-8450

Yoshiaki Kiyanagi
Graduate School of Engineering, Nagoya Univerity, Furo-cho, Chikusa-ku, Nagoya, 464-8601, Japan

Mamoru Fujiwara
Research Center for Nuclear Physics (RCNP), Osaka University 10-1 Mihogaoka, Ibaraki Osaka 567-0047 JAPAN

Hideki Namba
Executive Director, Japan Atomic Energy Agency (JAEA), 2-4 Shirane, Shirakata, Tokai-mura, Naka-gun, Ibaraki 319-1195, Japan

Hideto En'yo
Nishina Center for Accelerator-Based Science, RIKEN, 2-1, Hirosawa, Wako, Saitama 351-0198

Yoji Inui
Business Promotion Planning Department, Tokuyama Corporation, Kasumigaseki Common Gate West Tower 2-1, Kasumigaseki 3-chome, Chiyoda-ku, Tokyo 100-8983

Printed in the United States
By Bookmasters